Trim

CALGARY PUBLIC LIBRARY

NOV 2015

A Homeowner's Guide

STANLEY®

Trim

Steve Cory

The Taunton Press

Text © 2015 The Taunton Press, Inc.
Photographs © 2015 The Taunton Press, Inc. (except where noted)
Illustrations© 2015 The Taunton Press, Inc.

All rights reserved.

The Taunton Press, Inc., 63 South Main Street
PO Box 5506, Newtown, CT 06470-5506
Email:tp@taunton.com

Editor: Peter Chapman
Copy Editor: Seth Reichgott
Indexer: Jim Curtis
Cover and Interior Design: Stacy Wakefield Forte
Layout: Stacy Wakefield Forte
Photographer: Steve Cory and Diane Slavik (except where noted)

The following names/manufacturers appearing in *Trim* are trademarks:
Speed Square®, Surform®

Library of Congress Cataloging-in-Publication Data

Cory, Steve.
Stanley trim : a homeowner's guide / Steve Cory.
pages cm
Includes index.
ISBN 978-1-62710-942-0
1. Trim carpentry--Amateurs' manuals. I. Stanley Black & Decker Inc.
II. Title.
TH5695.C67 2015
694--dc23
2015022881

Printed in the United States of America
10 9 8 7 6 5 4 3 2 1

About Your Safety: Construction is inherently dangerous. Using hand or power tools improperly or ignoring safety practices can lead to permanent injury or even death. For safety, use caution, care, and good judgment when following the procedures described in this book. The publisher and Stanley cannot assume responsibility for any damage to property or injury to persons as a result of misuse of the information provided. Always follow all manufacturers' safety, installation, and operation warnings and instructions provided with the products and materials. Don't try to perform operations you learn about here (or elsewhere) unless you're certain they are safe for you. The projects in this book vary as to level of skill required, so some may not be appropriate for all do-it-yourselfers. If something about an operation doesn't feel right, don't do it, and instead, seek professional help. Remember to consult your local building department for information on building codes, permits, and other laws that may apply to your project.

STANLEY® and the STANLEY logo are trademarks of Stanley Black & Decker, Inc. or an affiliate thereof and are used under license.

MOLDINGS CAN BE AN EXUBERANT expression of personality, or they can just clean up the look of a room—and all for a modest materials cost and a day or so of labor.

Trim boards (the words "trim" and molding are used interchangeably) come in a dizzying array of shapes and sizes, and they can be finished with natural wood stain or paint. Many are inexpensive, allowing you to nicely trim out a room for under a hundred bucks. Trim boards can make a room feel satisfyingly finished and coordinated with the rest of the house's style, and they can be installed quickly and with a modest set of tools. However, if you have the budget and inclination, you can aim for a room with stunning details that pop out. In that case, you will need a few good tools, and you may want to study up to learn new skills—and spend some time practicing—before tackling the job.

This book helps you choose trim styles that suit your personal taste and your house's style, and it tells you all you need to know in order to cut and install moldings with tight joints and straight, crisp lines.

I'd like to offer special thanks to Mike Fish of Vogon Construction in Chicago for his sage advice and genial assistance. Danny Campana and William Shuman appear in many photos and built many of the projects. The folks at Owl Lumber in the Chicago area shared their vast expertise. And many thanks to the builders, interior designers, and architects who shared photos of their work, as well as the manufacturers who shared product photos.

CONTENTS

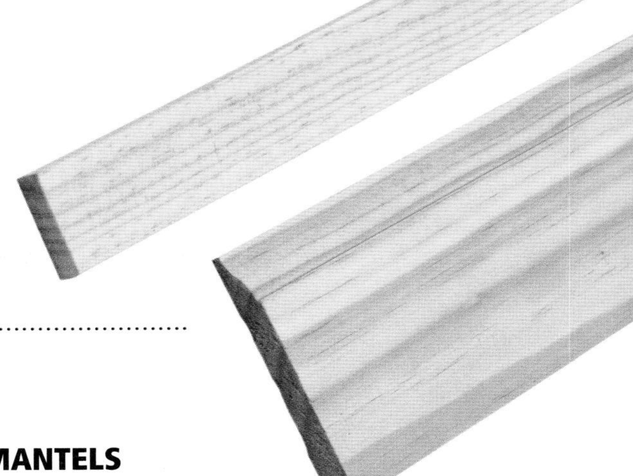

CHAPTER SEVEN

DOOR AND WINDOW CASING

CHAPTER EIGHT

CROWN MOLDING

CHAPTER NINE

WAINSCOTING AND WALL FRAMES

CHAPTER TEN

BEAMS AND MANTELS

CHAPTER ONE

WHAT TRIM CAN DO FOR YOUR HOME

ADDING TRIM IS LIKE PAINTING with broad strokes the outline of a room's features, strokes that add a satisfying sense of definition and dignity to the space. It makes a room feel finished.

Even in a home with neatly installed trim, upgrading the moldings can upgrade the look of the rooms. Trim has always had the virtue of covering a multitude of sins.

The details of wood trim shine through and add appeal. They also help establish a sense of style. Getting the trim style that you like—one that works with your taste in furnishings and your general design sense for your home—will help tie the rooms of your home together in a way that is satisfying and that makes your home feel like a welcoming space.

Adding additional trim pieces to your home, or changing the trim you currently have to a style you like better, is an upgrade that can make a big difference in the look and feel of your home's interior. If you do the work yourself, the expense will be the cost of materials, plus a few hundred dollars for tools if you don't already have them.

Trimwork does not call for special talents; the skills you need are taught in this book. However, every joint needs to be tight, and you must work carefully and systematically. If you have a devotion to detail and the patience to treat every board with care, you can achieve professional-looking results.

What Trim Is

If you've browsed the aisles of a lumberyard or home center looking at trim pieces, you know there is a lot to choose from, and it can appear more complicated than it really is. A room with basic trim will usually have just two types: baseboard, the trim covering the gap between the wall and floor; and casing, the trim that runs around doors and windows. A room with more elaborate trim builds up the baseboard and casing with additional profile pieces, and also adds trim elements elsewhere in the room. The array of trim profiles may be dizzying, but basic trim components are fairly simple.

A Rough beams and stone pillars blend with traditional trim for a charming rustic effect. The simple style of the pillar at the bottom of the stairs and the wood door suggest the Arts and Crafts style. B Called French country style by the designer, this home has Old World elegance reminiscent of a chateau. The bright open space employs relaxed color tones with broad elegant lines in the trim. Curved moldings like these can be easily achieved using modern materials. C The Farmhouse style in this home has traditional-looking trim with clean lines. Wide base molding and chair rail, trim pieces associated with traditional homes, have understated elegance in a casual setting.

A

B

C

D Elaborate trim like this may require a trip to an online store; in some cases detailed pieces may be made of polyurethane or other man-made products. **E** In a room with wallpaper, trim ties it all together with a neat bow. This powder room with satin white trim has a stylish, finished feel. **F** Homes with a modern décor often use simple moldings such as ranch casing and flat baseboards.

Floor-level terms

Baseboard covers the gap between flooring materials and the wall. In newer homes baseboard is often 3 in. to 5 in. wide; in more traditional homes the baseboard is often several inches taller and may combine several profile pieces.

Shoe molding is a small rounded piece nailed to the bottom of the baseboard to cover the gap between the baseboard and the floor.

A *plinth* is a decorative block at the bottom of door casing. Baseboard runs into the plinth, creating a seamless look between the baseboard and the trim.

Door and window terms

Casing covers the gaps around windows and doors. Wood jambs frame the raw openings for doors or windows, and casing covers the gap between the jambs and the wall material. If the casing goes all the way

A With tight joints and a good paint job, simple inexpensive Colonial casing can look great. B The three-part base in this room has simple modern lines that work well with other more detailed molding pieces; the overall effect is traditional yet crisp and straightforward looking.

around a window, it is *picture-frame casing*.

In older homes, a window often has a shelf-like piece at the bottom, called a *stool,* and an *apron* piece underneath to cover the gap between the stool and the wall.

Some casings use corner blocks or rosettes at the corners, for an old-fashioned look. When blocks are used, the casing does not need to be miter-cut.

Ceiling terms

Crown molding, also known as *cornice molding,* covers the gap between ceiling and wall. Crown may be made of a single wide molding board. For a richer appearance, crown molding may combine several layers of profiles for a more substantial effect.

Beams are structural pieces, usually covered with drywall. When exposed, they may be trimmed with molding. A faux beam is just what it sounds like: beamlike in appearance, but with no structural function.

In ceilings with beams running in two directions, the recessed spaces between beams are called *coffers*.Traditionally, coffers were filled with plaster decoration. But they may also be filled with a plain or detailed wood surface, and molding may be installed along joint lines between beams and coffers.

C The high ceilings in this home accommodate a double band of crown molding, making the beams and coffers look like a lavishly framed work of art. D The simple homemade design of the casing around this bathroom storage closet has Craftsman charm, whereas the trim detail below the window suggests the Art Deco period; historically, the two periods overlapped. E This window has mitered picture-frame casing, which means that the stool is narrow. There's no room for a potted plant, but a flatter window profile does give the room a more spacious, open feel.

Wall terms

Traditionally, walls were made from plaster, and the damage from wall hangings or furniture bumps was a big nuisance; many wall trim applications evolved to prevent plaster damage.

Chair rail is a horizontal band of trim that runs along a wall at the same height as a chair back, originally meant to protect walls from damage from chair bumping. Your home center may not carry a wide selection, but other trim pieces can be used singly or in combination to create a chair-rail effect.

Wainscoting is wood paneling—beadboard is one option—that attaches to the lower part of the wall, visually dividing the room into a top and bottom section.

Wainscot cap goes on top of wainscoting and is similar to chair rail.

Plate rail can go atop wainscoting or run horizontally around the room on its own like chair rail, only higher up at eye level. It is a little wider and traditionally served as a mini-shelf to display prized Limoges china or other artwork.

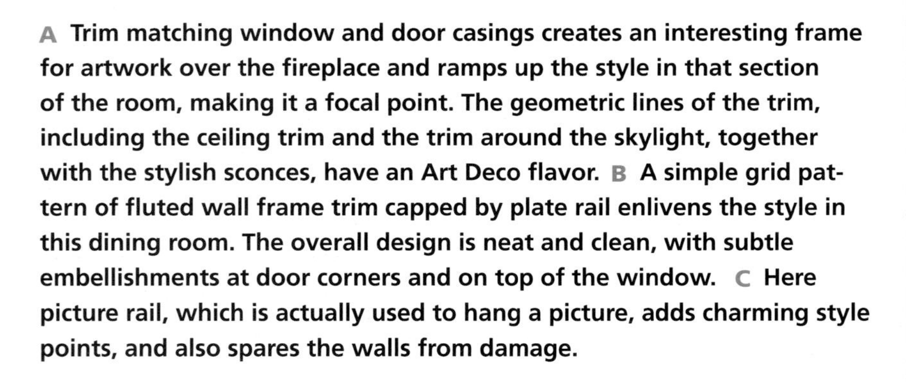

A Trim matching window and door casings creates an interesting frame for artwork over the fireplace and ramps up the style in that section of the room, making it a focal point. The geometric lines of the trim, including the ceiling trim and the trim around the skylight, together with the stylish sconces, have an Art Deco flavor. B A simple grid pattern of fluted wall frame trim capped by plate rail enlivens the style in this dining room. The overall design is neat and clean, with subtle embellishments at door corners and on top of the window. C Here picture rail, which is actually used to hang a picture, adds charming style points, and also spares the walls from damage.

Picture rail also runs around the room horizontally, but higher still, for hanging pictures or paintings. It has a profile that extends out to accept hooks for hanging frames. In some older homes it doubles as crown molding, often in smaller rooms.

Wall frames are made from panels and trim pieces to form rectangular shapes in open spaces on walls, usually in homes with a traditional style.

Pilasters are flat pieces, reminiscent of ancient Greek or Roman pillars, that attach vertically to a wall, often to showcase a focal point in a room. For example, they may be placed on either side of a fireplace or a built-in section of shelves.

Stair terms

A basic set of stairs usually has four components: *treads,* the flat pieces you step on; *stringers,* the notched side pieces that the treads nest into; *risers,* the vertical pieces that make a right angle to the treads (metal stairs usually do not have these); and the *handrail,* if needed. Adding trim makes stairs a stylish player in the home's overall appearance.

Balusters, also called *spindles*, are common in older stairs. They are often installed two or three per tread and connect to the handrail at the top. In more modern homes, balusters may run parallel to the handrail. Together the row of balusters is called a *balustrade*.

Newel posts anchor the balustrade at the top and bottom. The top section of the post is frequently decorated with trim, often featuring layers of molding both on top and sometimes along the sides as well.

D Natural wood in the balustrade of these handsome traditional stairs makes an appealing contrast to the clean white stringers and sleek black treads. Trim along the edges of the stringers matches detail in the home's door casings and baseboard. The newel post with its rich traditional detail is the star of the show. (Courtesy of Schwartz and Architecture, Matthew Millman Photographer.) E Beaded wainscoting extends higher than usual in this bathroom with Cape Cod ambience. F Picture rail beneath ceiling curves seems to underscore the beauty of old-style plaster work.

Finding a Balance

Trim is an architectural feature of a home and, as such, it needs to make sense with the home's overall style and proportions. The key word is balance. There are a few concepts to keep in mind to ensure your home's trim does not look awkward or out of kilter.

Proportion

Trim sizing should fit the size of the room. The ancient Greeks and Romans developed standards called "orders" that architects still use today. Orders describe proportions for pillars (like the stone pillars holding up the Acropolis), which were broken down into three main sections: a base pedestal, a long column, and a top section called an entablature. In terms of interior trim, these sections correspond to base molding, wall height, and crown molding.

For instance, the Greek Doric order, often adopted by the Craftsman style, has columns that are somewhat short and squat. It is calculated as follows: Pedestals at base are 2¾ parts tall (corresponding to baseboard); columns are 8 parts tall; and entablatures (corresponding to crown molding) are 2 parts tall. In all, this system has 12¾ parts, so divide your room's height by 12.75 to determine how long each part is for your room.

Victorian trim tended to be based on the Corinthian order, which has a taller entablature section and looks less substantial and more graceful.

A pleasant balance

Here are some other basic rules of thumb to help avoid an unbalanced-looking overall appearance:

- Molding at the top and bottom of a room should flare out from the wall (that is, become a bit thicker) to look as though it is able to support the load (which is what the top and bottom parts of a column did).
- Window and door casings should usually be consistent with crown and base molding. If trim is wide at the base and ceiling of a room, it may look odd if trim is narrow by doors and windows (especially at the top).
- If you have trim with prominent detail serving as a cap piece—for instance, at the top of a door—the flat piece underneath should be taller to look like a good fit.
- Maintain a level of consistency throughout the house. Spacious rooms may have trim that is more ornate, whereas smaller rooms can have the same basic style with fewer build-on pieces. Or wide crown molding in a spacious room may be replaced with a thin piece of trim, like picture rail, that matches other trim elements in a smaller room.

A A home with high ceilings can accommodate broad trim pieces. In this case, contrasting wall colors together with wood tones highlight the details of the trim, which is balanced and feels like a good fit. B The neat pencil trim in this office dresses up a plain modern space and makes it more interesting and inviting. Because this room has high ceilings, it could accommodate wider trim, but color and trim choices divide the room into smaller sections and make the narrow trim fit.

C Classic dentil crown molding, popular in the Georgian period (in the 1700s), offers an interesting texture with alternating shadows, its detail nicely sized to be appreciated from the distance of eye level. D This Craftsman-style home features proportions that follow the Doric order. The space is well proportioned and more compact than the airy graceful effect of a Victorian room, which with higher ceilings typically follows proportions established by the Corinthian order. E Wood trim that plays off the wood tones of the flooring brightens up the style in this modern home. A simpler use of trim in the family room throws the spotlight on the piano, whereas additional pieces in the dining room give the room a formal feel, yet both rooms flow together smoothly.

Choosing Trim Shapes

If you've walked through a lumberyard looking at trim pieces, it may seem like a complicated maze to navigate. But the maze is simplified when you sort the pieces in two basic categories: (1) straight lengths that may be narrow or wide; and (2) pieces that either curve in toward the wall or out toward the room—or both.

Trim looks different depending on your vantage point. Crown molding and other trim that's higher than eye level tends to look better if it curves down toward the viewer; on the other hand, base trim, which is lower than eye level, tends to look best if it curves up toward eye level. When choosing trim, it's great if you can look at it from a height that approximates the installation.

Here are some other guidelines:

- Aim for balance with shadows. The nooks and crannies in trim pieces when viewed from below or above will have shadow lines that make trim more interesting. Curved pieces have softer shadows, whereas flat sections or tight grooves have darker, more pronounced shadow lines. A combination of the two tends to be more visually appealing.
- Add some variety. When you're combining trim pieces in one section, perhaps for crown molding, experiment with sizes and designs. For instance, you might combine a large cove and a small cove, or combine pieces that curve in opposite directions, separating curves with a thinner bead piece. If you want to use "carved or pressed-in" designs, pair them with plainer molding pieces to better showcase the details.
- Repeat themes for a unified look. Trim around windows and doors usually should share the same molding pieces and joinery. If you want to boost style in one room, vary the theme by adding an extra piece of molding, at the top of a door, for instance.
- Consider sight lines. Fine detail in trim that's not at eye level may be hard to see, so a room's most intricate detail is best positioned where it will be viewed straight on—perhaps along a fireplace, at the top of a built-in bench, or on the side of a stairway. When trim is farther away, the detail needs to be bigger to be pleasantly noticeable.

A The crown molding, fireplace mantel, and tops of the door and window in this room all share the same flared trim profile.

Matching Style in an Older Home

If trips to the lumberyard or home center have failed to yield an exact match to your home's molding among the array of existing pieces, you may be tempted to pay for expensive custom millwork. But a small difference in shape is barely noticeable, as long as trim is consistent within a room and the design is similar in neighboring rooms. Of course, you should try to match the spirit of the style, but if the details vary a little bit, it will be a minor variation on the home's style theme, and the overall effect will still be nicely unified.

B In this room the chair rail has the most ornate trim profile, echoed in a scaled-down form in the window casing. Flatter baseboard is a pleasant contrast, and thin wall frames add a formal touch that softens the modern style of this living room. C The entryway casing trim nicely coordinates with the fireplace mantel: Both feature profile detail with geometric lines that are simple, stylish, and unusual—true to the Art Deco theme. (Courtesy of Schwartz and Architecture, Matthew Millman Photographer.) D A trim profile that combines curved and flat pieces produces shadow lines that make the effect visually interesting. The ceiling is a distance away, but the detail in the crown molding, which flares out at the top, is pleasantly balanced so a person standing below can appreciate it. E This interesting trim has raised keystone sections at the center above doors, a detail that was popular in Federal architectural style of the late 1700s and early 1800s. The trim at the top alternates flat and curved sections with beading in between for a richly decorative effect.

A Short History of Trim

Trim styles, like other matters involving personal taste, have evolved over time. Certain styles were all the rage for a number of years and then faded in popularity. For instance, during the years of Queen Victoria's reign, style preferences trended toward ornate detail, in trim as well as in furnishings and clothing. But after a few decades of this, people grew tired of lavish, over-the-top decorating that required materials imported from abroad, and were more interested in simple, honest, home-grown decoration. The Arts and Crafts movement (Frank Lloyd Wright was one of its stars) became a hot new trend beginning in the 1890s. As time went on, the Art Deco style became popular.

In the 21st century, style preferences are all over the map, embracing the old along with the new. Traditional homes have an enduring appeal that feels welcoming to many people. A more casual style, perhaps best termed "cottage," has become popular in recent years—based on its dominance in home decorating catalogs that showcase products in settings like a summer beach cottage, a farmhouse, or an old European country home. Trim in

A Abundant Classical-style trim, with a special emphasis on fluted pillars, is reminiscent of the Greek Revival period, which was popular in the late 18th and early 19th centuries, when interest in archeological discoveries awakened a taste for all things Greek. B Colonial-style homes often had chair rail to protect plaster walls (though wainscoting came later), and fireplace surrounds were common. Baseboards were substantial to keep out drafts and included shoe molding. Crown molding had a similar profile to window and door casing.

C Wall frames, common in Colonial-style homes, were often used to dress up living rooms and dining rooms. In this newer home, lavish wall frames that pop out against a backdrop of blue wall color achieve a whimsical effect. D Classic pillars abound in this over-the-top Victorian bathroom. In the foreground they seem to be weight bearing, whereas those in the background are pilasters—flat trim pieces resembling pillars. Pilasters are often used to frame focal points in a room, placed on either side of a doorway, fireplace, bed, or here, a sink.

these homes may be simpler and may appear worn or handmade, and is often a pared-down form of traditional.

Colonial

In the earliest days of the country's existence, before 1776, most homes were simply built by hand by recent immigrants and tended to reflect styles of the old country, with most people coming from England, France, and Spain. Most Colonial-style homes now in existence were built about 100 years after the Declaration of Independence during a patriotic period and are actually from the Colonial Revival period. This period overlapped with the Victorian era, when wood from the West was abundant. The hand-planed millwork for Colonial-style trim was a little fancier than homes built a century earlier, but not as ornate as Victorian trim. Early trim featured mainly curves and flat sections, both of which were easy to do with hand-planing tools.

Classical

In the days following the American Revolution, tastes evolved toward classical lines, borrowing from ancient Greek and Roman styles. There were a number of variations on this classical theme, including the Georgian, Federal, and Greek Revival architectural styles. The trim at the top and bottom of ancient columns provided inspiration for the lines and proportions used to establish new variations on old themes.

Victorian

During the Victorian era, fancier was better. Furniture styles were gaudy and over-the-top and soon faded in popularity, but the stylish trim of this era has enduring appeal.

Arts and Crafts or Craftsman

This style originated in England at the end of the 19th century as the Arts and Crafts movement, and then migrated to the United States, where it has also been called the Craftsman style. It was a movement toward simpler details with an emphasis on local materials—fir and oak wood—and workmanship. It was a movement away from imported trim with lots of ornamentation. Unpainted wood tones were popular, but when trim was painted, the colors tended to be white or soft shades of green, like sage or olive. Craftsman style was easier for the average homeowner to execute, as the joinery was not as tricky.

Art Deco and Modern

Art Deco in the 20s and 30s and Modern in the 50s and 60s were trim styles that trended toward simple, clean lines. The Art Deco style featured angular and geometric lines with a simple but stylish and sometimes fanciful flair. The Mid-Century style minimized trim, but generally didn't eliminate it because of the demanding level of skill needed to achieve perfect joints without trim to cover imperfections. With Modern styles the overall emphasis was more on the home's textures, colors, and materials.

Nowadays it's common for a home not to have a style name. People tend to go with what they like, and as long as you work within the parameters of your home's style and the proportions of its rooms, there is a lot of leeway to create the look that feels right for your home.

A Delicate balusters like these are common in Victorian-style homes. The contrast between wood tones and painted trim gives the stairs a crisp, formal feel.

B In Victorian homes, high ceilings were the norm, and trim became wider and more detailed. The graceful, substantial trim in this living room seems well proportioned to the dimensions of the room. C Simple casing around doors and windows was the trend during the Arts and Crafts period. Fireplace mantels were plainer, too, and tended to feature a simple shelf. If wood was not natural tones, it was often painted white or green. D Natural wood, like the oak trim here, was popular during the Arts and Crafts period; built-in bookcases were frequent additions. Arts and Crafts moldings feature wide flat spaces for a simple, straightforward look. E Modern trim is usually fairly simple and straightforward. In this bathroom, the picture-frame window trim has a traditional feel, which plays off the sink design, whereas the rest of the room, except for the baseboard trim, has clean lines with no trim.

CHAPTER TWO

TRIM MATERIALS

VISIT A HOME CENTER and you'll find a large assortment of trim moldings (a word that is often spelled "mouldings," perhaps to make it seem more high-toned). Where moldings once were either pine or hardwood, now you can find trim boards that are prefinished, medium-density fiberboard (MDF), polyvinyl chloride (PVC), polystyrene, vinyl, and vinyl-clad. And scrolling through online sources will reveal an even more dizzying array of materials and decorative options. Many of these products are less expensive than solid wood, whereas others do not need to be stained or painted and may offer superior durability. This chapter helps you choose the best products for your situation and your budget.

It was once common to say that applying trim is a cheap way to decorate a room. That is not as obviously true as it once was, because trim prices have risen dramatically in recent decades. So it's more important than ever to select cost-effective materials and to plan the job so as to avoid waste.

Trim can certainly be applied with a hammer and finishing nails, but nowadays power nails are a popular choice even among homeowners. In some situations, finish screws are also a good option.

In addition to the many trim profile options, you can assemble two or more standard pieces to produce a trim profile of your own. If you are ambitious, you may even want to mill your own trim using a router table.

Molding Shapes

First, choose the molding profiles, or shapes, for various parts of the room. Most profiles are available in hardwood, softwood, and a variety of manmade materials. Let's start by learning some basic shapes and types.

Molding profiles are made of simple shapes, often built up in combination. A *cove* is an inward scallop, whereas a *bead* protrudes outward in a half circle. A *thumbnail bead* is a half-circle that dead-ends into the wall. *Ogee* refers to a number of profiles that contain an S-shape; the top and the bottom of the S are typically parallel. A *fillet* is a small right-angle cutout. A *quarter round* is exactly what you'd expect; an *ovolo* is a quarter round with a fillet at the top and bottom.

TIP **The term *sprung molding* refers to molding that has two beveled edges so it can mount to two surfaces at right angles to each other (most commonly, a wall and ceiling). There is a triangular empty space behind sprung molding.**

BUILT-UP MOLDING STACKED ON EDGE

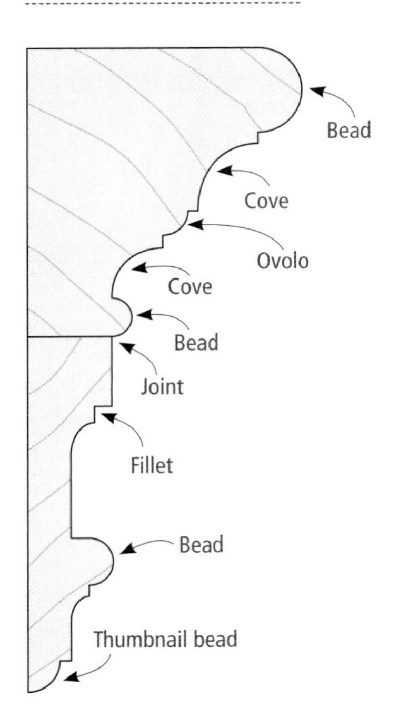

BASIC PROFILES

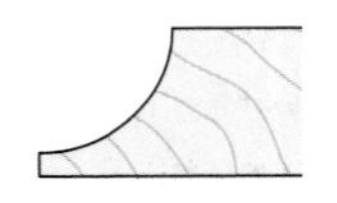
Cove

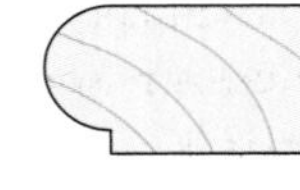
Bead

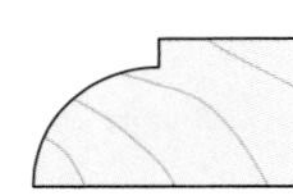
Thumbnail bead

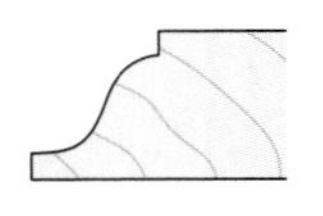
Ogee

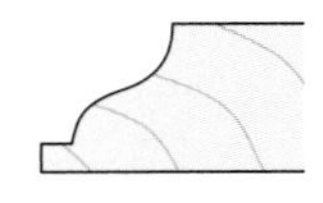
Reverse ogee

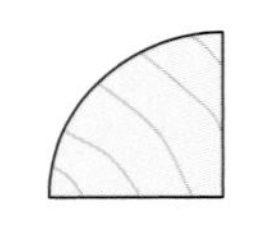
Quarter round

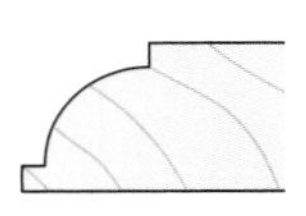
Ovolo

Four Basic Styles

The large number of molding shapes can be divided into four basic categories, typified by the casing profiles shown here. A *clamshell* or *Ranch* style is a simple curved edge without coves, fillets, or beads. It is often used in modern-style homes. *Colonial* moldings typically have fairly complicated profiles, for a richer appearance. *Fluted* trim has a symmetrical pattern of beads, for a stately appearance. And *Arts and Crafts*-type moldings are plain, with few or no flourishes.

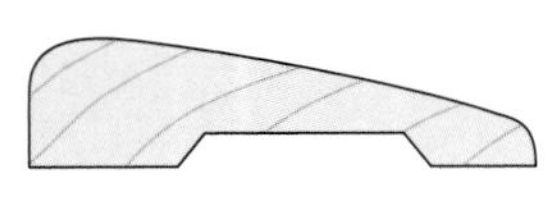
Clamshell

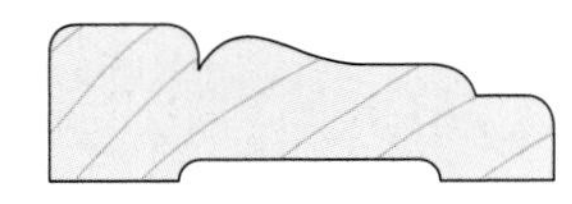
Colonial

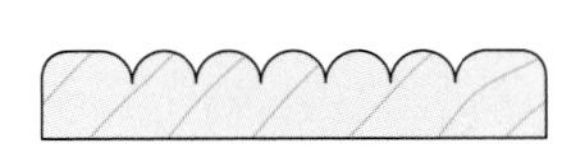
Fluted

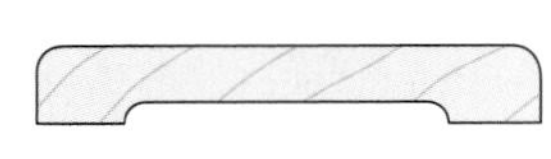
Arts and Crafts

Decorative Moldings

Trim boards that are a bit out of the ordinary may be available in a special aisle in a home center or lumberyard, and are certainly available from online sources. Some pieces are reasonably priced, while others may cost over $100 for a single 10-footer.

Many of the moldings shown here may be described as casings, and can be used for that purpose. But they can also be used for other decorative treatments, such as built-up crowns.

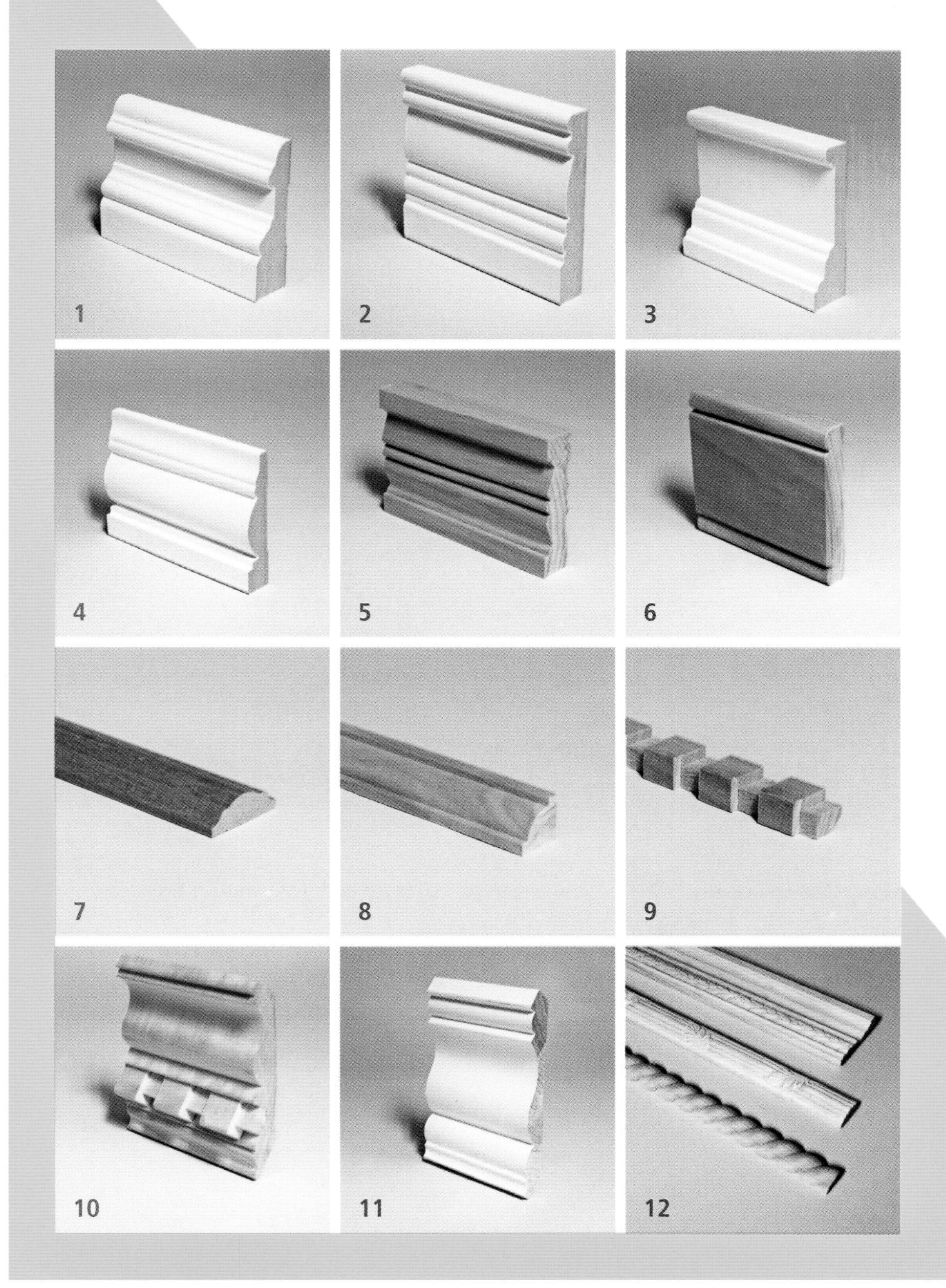

1. Base

2-4. Casings

5. Wide head casing

6. Chair rail

7. Mini-mold

8. Reverse cove

9. Dentil insert

10. Crown with dentil detail

11. Wide crown

12. Molded or carved trim pieces. You can choose among a huge variety of moldings like these, with carved decorative features. These are made of natural wood, but others are molded from PVC or other materials. They are generally pretty expensive, but you may want to use one or more for a feature such as a fireplace surround or to liven up some shelves or cabinets.

STANDARD MOLDINGS

Shown here are some of the most common moldings. In most cases, you can find them in pine, oak, primed MDF or primed pine, vinyl, and most of the other materials discussed on pp. 24–25.

1

CASING

1. Ranch or clamshell casing
2. Colonial casing
3. Fluted casing
4. Back band
5. Plain or round-edge casing

2

4

3

5

1

2

BASE AND BASE SHOE

1. Colonial base
2. Quarter-round
3. Base shoe
4. Base cap

3

4

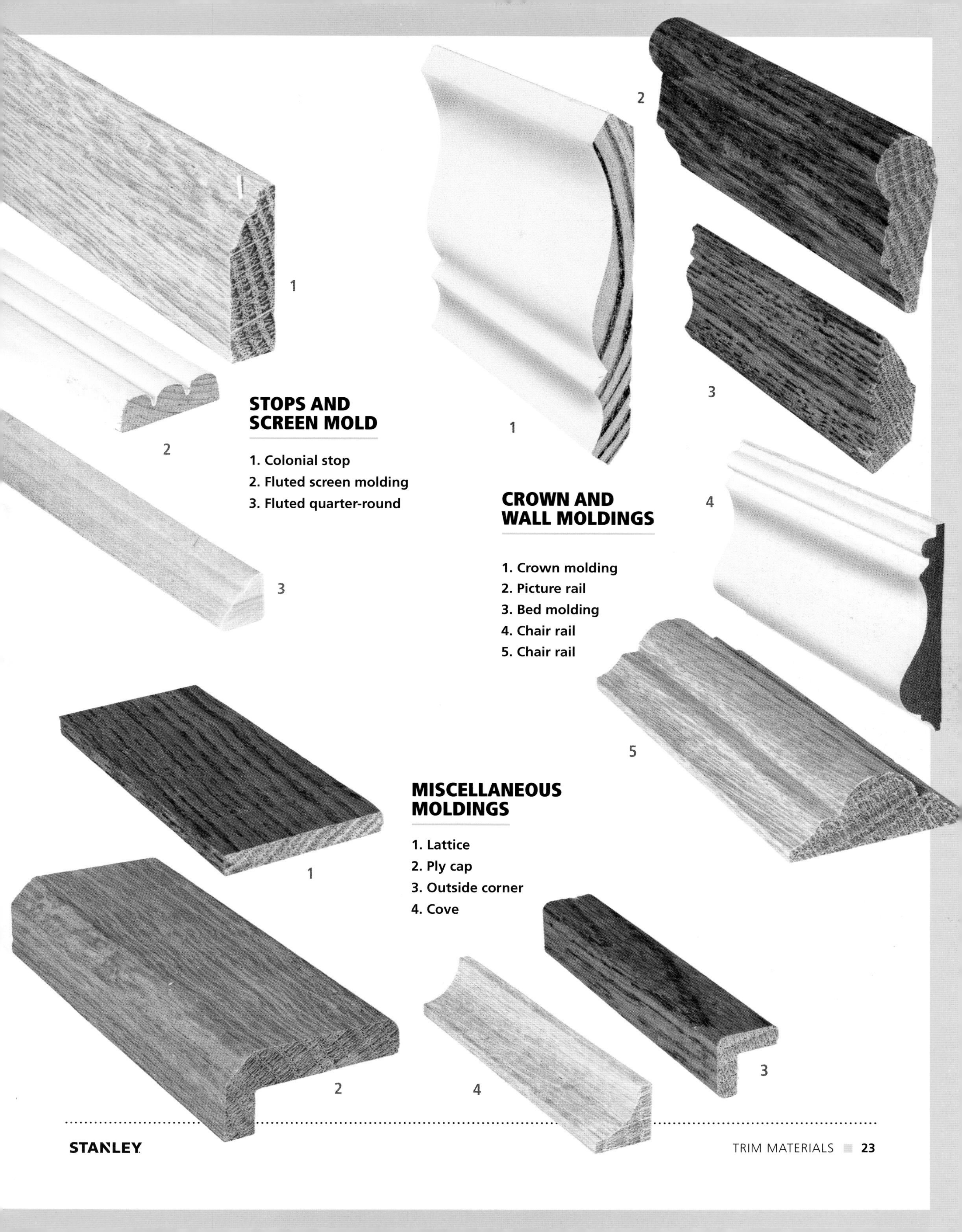

STOPS AND SCREEN MOLD

1. Colonial stop
2. Fluted screen molding
3. Fluted quarter-round

CROWN AND WALL MOLDINGS

1. Crown molding
2. Picture rail
3. Bed molding
4. Chair rail
5. Chair rail

MISCELLANEOUS MOLDINGS

1. Lattice
2. Ply cap
3. Outside corner
4. Cove

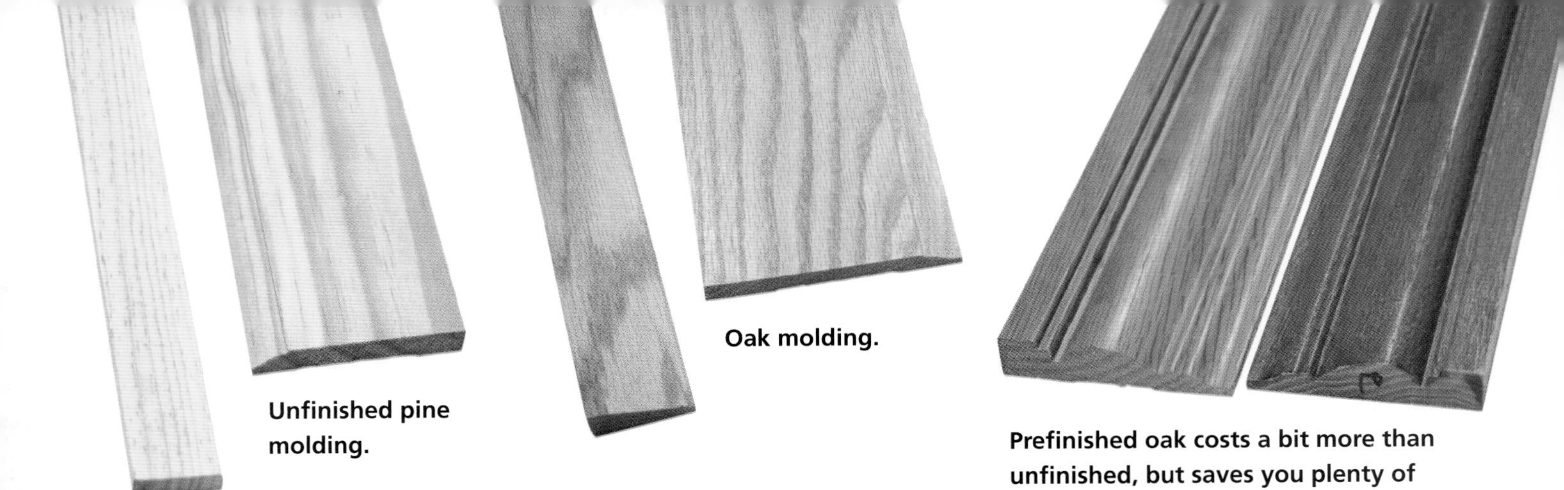

Unfinished pine molding.

Oak molding.

Prefinished oak costs a bit more than unfinished, but saves you plenty of time and hassle. Many suppliers carry two colors, natural and dark.

Wood and Other Materials

When shopping for moldings, you'll see wood trim boards described by their species and MDF labeled as such. But synthetic moldings often are not listed by their actual material; you may need to read the literature closely to tell if it is polystyrene, vinyl, or PVC, for instance. Home centers and lumberyards carry different types, but here are some of the most common.

Unfinished softwood

Softwood trim is often made of pine but is sometimes fir or hemlock. Softwood is light and easy to cut and fasten, and has a natural handsome appeal. It is common to paint softwood, but careful staining and finishing can produce a rich appearance. Softwood is strong and fairly resistant to moderate exposure to moisture, but it can be dented.

Hardwood

Oak is the most popular and readily available type of hardwood molding, but if you've got some extra bucks you can go to a hardwood supplier and get maple, cherry, or even hickory. Hardwood trim is somewhat more difficult to cut and fasten than softwood, but if you prefer a natural or stained wood finish, it will give you a beautiful result, and it is difficult to dent as well.

Primed or painted

Painting can often take much longer than installing. So if you plan to paint the molding, consider buying primed or painted wood moldings. These moldings have the characteristics of softwood trim.

Straight and Wavy Grains

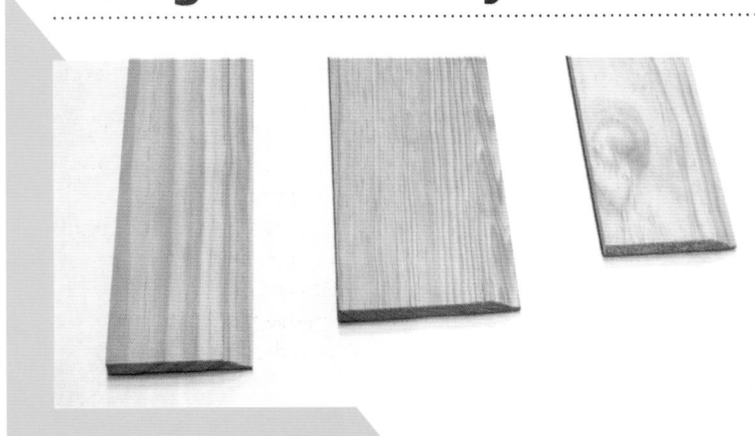

Trim boards with wide, wavy grain lines like the one on the right will most probably remain warp-free as long as they are fastened securely to the wall. But boards with narrow, straight grain lines (like the board in the middle) are more stable and certain to stay nice and straight. Many boards (like the one on the left) have a combination of straight and wavy grains. If you have a choice, choose as many straight-grained boards as possible.

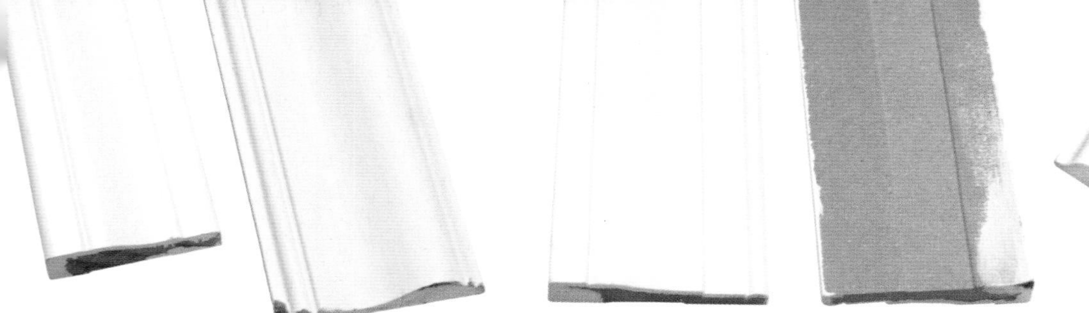

Vinyl trim.

Primed trim pieces like the one on the right are ready for a top coat; painted trim (left) needs only touching up, as long as you want it to be white.

MDF trim.

Polystyrene moldings.

MDF trim

Medium-density fiberboard is heavier than natural wood, but it cuts and fastens fairly easily. Trim made of MDF comes with a solid coat of paint, and it's often the least expensive option. MDF also resists denting. However, if it gets even a little wet, MDF will swell unattractively. So use it in rooms that are not likely to be moist, and apply primer to the back side before installing.

Polyurethane, vinyl, PVC

Moldings made with strong plastics like cellular vinyl, polyurethane, PVC, or even resin can be expensive, but they are easy to work with, strong, and completely immune to rot. Be sure the type you choose is paintable, and pay attention to the type of paint that is recommended. Often, a 100-percent acrylic paint is a good choice.

Polystyrene

Many home centers carry a line or two of inexpensive moldings that are very lightweight. Some have a printed faux-wood design, whereas others may be a solid color. Though the label may not say so, these are made of a plastic foam called polystyrene. These may be a good option in certain spaces, but be aware that they dent easily, and cutting or fastening imperfections cannot be hidden with caulk. Polystyrene should be installed with a nail gun; using a hammer with finish nails will lead to an unattractive installation. Painting this product can be difficult; you may need to apply an alcohol- or oil-based primer first.

TIP **If you want to stain your molding, resist the temptation to buy "paint grade" trim. It may look like you can stain it, but the results will be far less attractive than what you can achieve with "stain grade" molding.**

Finger-Jointed Trim

Finger-jointed trim is made of short pieces solidly glued together with intricate jointing. (Newer finger joints appear flat, whereas older finger joints looked more like fingers.) Finger-jointing is a "green" option because it makes the most efficient use of available lumber. Protect finger-jointed boards with at least two good coats of paint, because the joints may delaminate if exposed to moisture.

Blocks and Rosettes

To add interest and make the installation easier, consider installing blocks at corners for base molding and rosettes for window and door casing. Special blocks can even be used for crown molding. Once blocks are installed, you need only cut the moldings at right angles, rather than making difficult miter, bevel, or coped cuts. See pp. 94–95 and 130–131.

Check for cup.

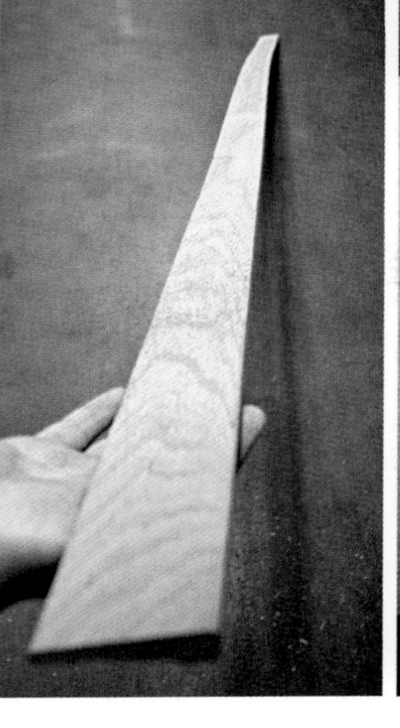

Check for bow.

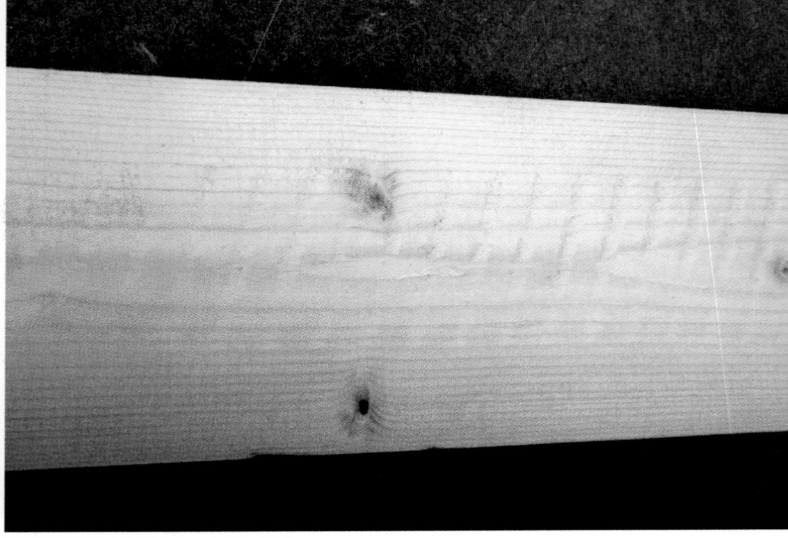

Check for mill marks.

Choosing boards

Synthetic trim boards are all pretty much the same, but wood boards can vary in quality. Because the boards are thin, minor curves and twists can be straightened out during installation, but pieces with severe distortion may crack during installation.

If a board is cupped—bowl-shaped along its width—try bending the board with your hands to straighten it. If you cannot, choose another board.

Sight down the length of a board. If it deflects more than ½ in. over the length of 8 ft., choose another board.

Some boards have indentations or a series of wavy marks left by the milling process. These will likely become more pronounced once the board is painted or stained. You could sand them out, but it is probably easiest just to choose another board.

Beadboard

Wainscoting is often made using beadboard, which is milled with decorative vertical lines. This is available in several ways. Individual pieces (above left), with a groove down the middle, fit together via a protruding tongue on one side and a receiving groove on the other, so the pieces nest together. There is a groove down the middle that mimics the appearance of a tongue-and-groove joint. Individual boards may be ¾ in. or ⅜ in. thick. You can also buy sheets that imitate the look of beadboard (below left). These are quick to install and, once painted, look much like real beadboard.

Fasteners

Trim is usually installed with nails rather than screws. For centuries trim has been attached using finish nails pounded into place with a hammer, and that method is still used today. However, nowadays most pros and many homeowners use pneumatic nail guns.

Hand-driven finish nails

Nails that you drive by hand are designated by their "penny" size; the higher the number, the larger the nail. Oddly enough, a box of hand nails will often signify the "penny" size with the letter "d," short for "denarius," an ancient coin.

3d (or 3-penny) nails are 1¼ in. long, and 4d nails are 1½ in. long. Both are used for attaching thin trim pieces like base shoe, door or window stop, and the thin edges of casing. 6d nails are 2 in. long and are the most common nails for general trim installation. At 2½ in. long, 8d nails are used where extra holding power is needed or where a nail must travel through thick finish wall material before meeting wood framing. 12d and 16d finish nails are rarely needed. "Hard-trim" nails are extra thin.

Panel nails come in colors, usually brown and white. They feature small heads and ribbed shanks for greater grabbing power. Brads are nails smaller than 3d.

Pneumatic nails

Nails driven with a power nail gun are designated by length, the type of head, and gauge (or "ga."); the smaller the gauge number, the thicker the nail. For the advantages of pneumatic nailing, see p. 43.

Pin nails are short, and a svelte 18 or 23 gauge. They don't hold very well by themselves because they don't have heads, but they can be driven into almost any piece without causing a split. They are called "pin" because you need to drive two of them at different angles to pin a board in place. Pin nails are often used to hold a piece temporarily until glue sets.

Other 18-gauge nails may be called brads. These can be anywhere from ½ in. to 1½ in. long, and have slight heads for a bit more holding power.

16-gauge finish nails are the most common nails in trim work. They have heads for holding power and range in length from 1 in. to 2½ in.

Trim screws

For even more holding power, or if you need to fasten in an awkward place and don't have a power nailer, consider finish screws, which are driven with a drill. To ensure against splitting, you'll need to bore pilot holes first.

POWER-DRIVEN FINISH NAILS

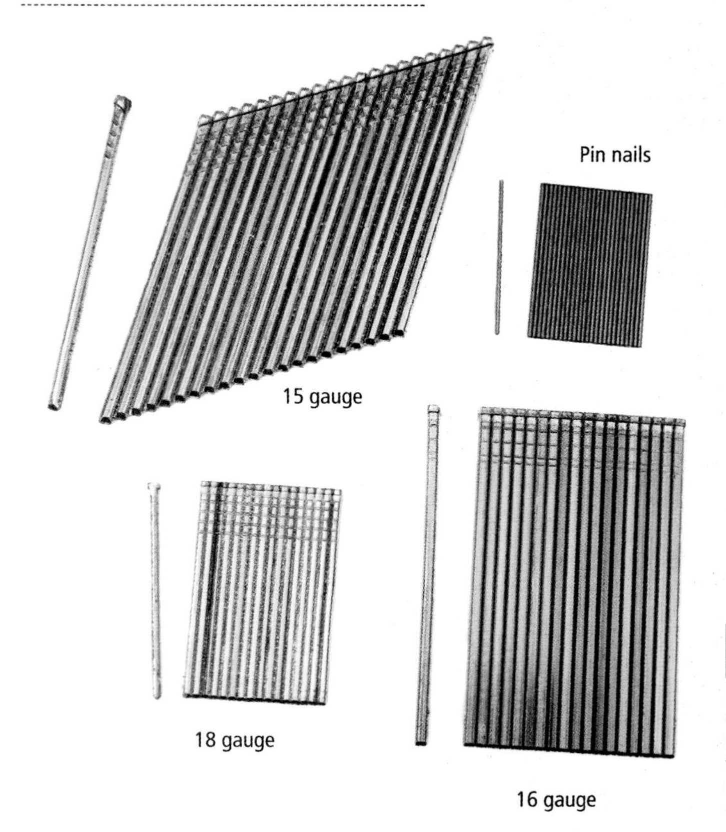

TRIM SCREWS AND DRIVER

TIP **Phillips-head trim screws are available, but it's distressingly easy to strip out the heads while driving. Square-drive screws are far less prone to stripping.**

TIP **Larger finish nails have more pronounced heads, for example, 15 gauge, and are up to 2½ in. long. Use these for installing heavy exterior doors, or wherever extra holding power is needed.**

Wood Glue

Wood glue is often used in trimwork to strengthen joints. For interior work, use standard PVA glue. If you're installing trim outside or in a place that could get wet, consider polyurethane glue, which swells and foams as it dries; you'll need to sand the surface of the joint.

CHAPTER THREE

TOOLS

IF YOU'VE GOT ONLY a few small trim jobs to perform, you can do them using a simple and very inexpensive set of hand tools: a hand miter saw, perhaps a coping saw, a measuring tape, a hammer, and nail sets. (A plane, knife, squares, pry bars, and chisels also come in handy.) And if you've got time on your hands, you can tackle most any trim job (other than wide crown molding) using those tools.

But more sophisticated tools make the job go a good deal more smoothly and quickly, and produce more reliable results. In particular, an accurate chopsaw and a set of pneumatic nail guns can more than halve your installation time and greatly reduce the chances of denting the wood.

In most cases, good-quality mid-priced tools are the best option. Very expensive power tools can be heavy and a bit clunky to operate for most modest operations. Unless you are a professional who will install elaborate trimwork on a regular basis for years, all the tools you need will cost less than $600 new.

Cutting Tools

For professional-looking results, use tools that will make clean and straight cuts. They should also be easy to use and adjust.

Miter saw

A power miter saw, sometimes called a chopsaw, has become a standard part of many wood shops, as well as a constant companion for many carpenters. It makes quick work of making precise and clean cuts in trim boards. Properly adjusted (see p. 56), it will consistently cut boards at exactly 90 or 45 degrees and will also cut perfect bevels.

A standard miter saw, sometimes referred to as a simple miter saw, can make simple miter cuts and crosscuts. A single compound miter saw, often just called a compound miter saw, can make miter cuts, bevel cuts, and crosscuts. It can tilt in one direction to make most compound cuts—the combinations of miter and bevel cuts—needed for crown molding. A dual compound miter saw, sometimes called a double-bevel miter saw, tilts in both directions, making it easier to figure out and make compound cuts. A sliding compound miter saw slides to cut wide boards, and can create compound miters as well.

Be sure your saw can cut the boards you will install, at all the angles and bevels you will use. For instance, a standard chopsaw with a 10-in. blade will easily handle almost every cutting task. A typical 10-incher can make a 90-degree crosscut that is 6 in. wide, as well as a 45-degree miter cut on a 4¼-in.-wide board and a bevel cut on a board 6 in. wide. A smaller saw will handle most baseboards and casings. If you will cut very wide crown moldings, you may need to buy a sliding compound miter saw, or a compound miter saw with a 12-in. blade.

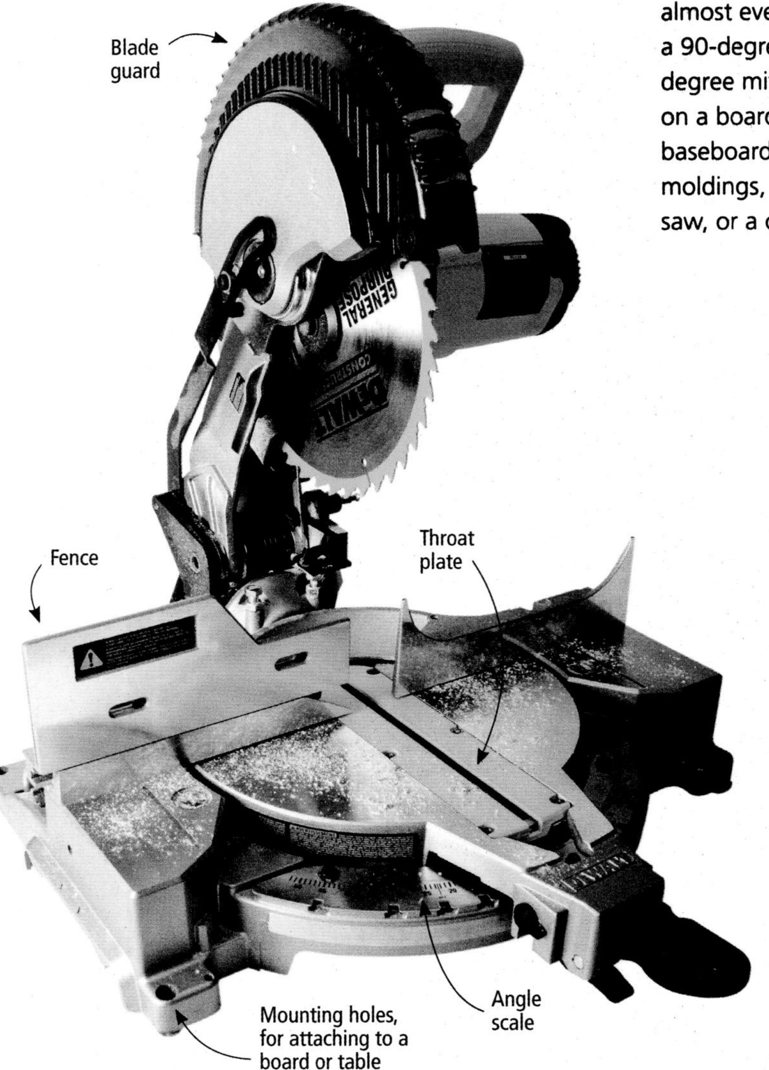

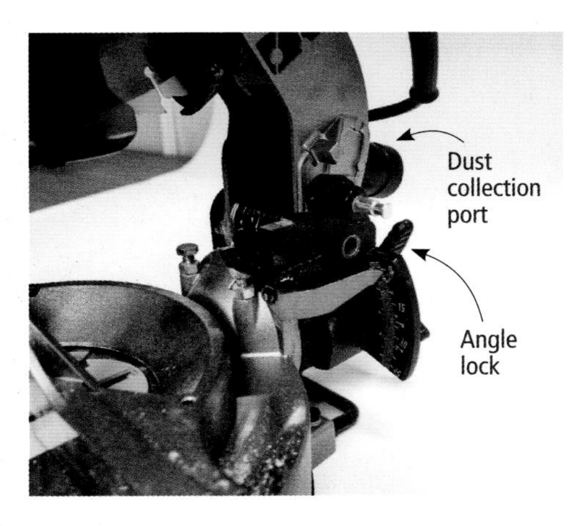

This 10-in. single compound miter saw is ideal for most trim projects, with the exception of very wide crown molding.

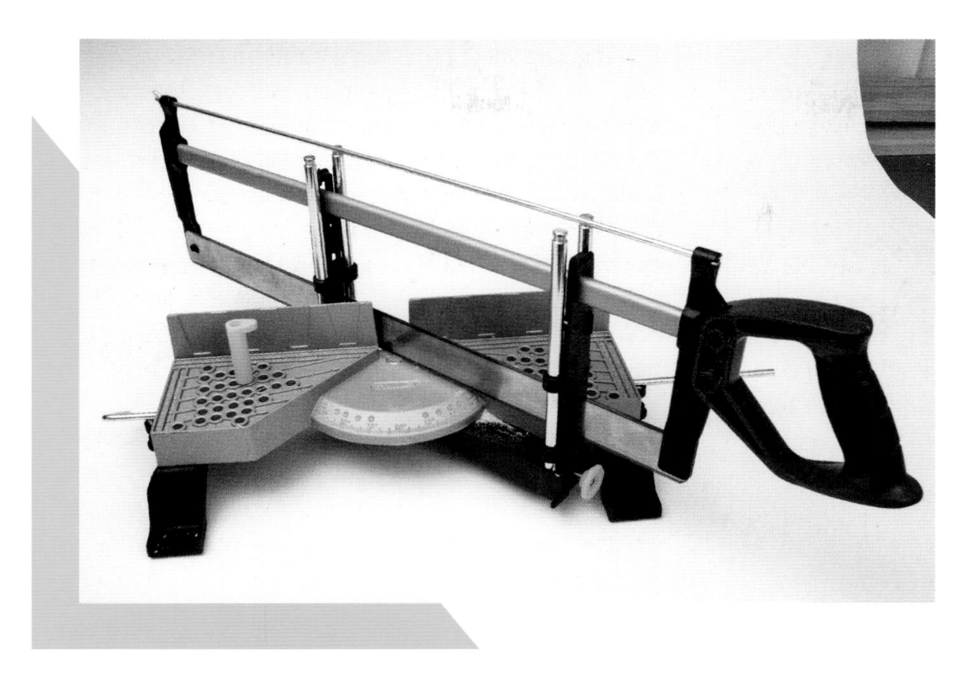

Hand Miter Saw

A hand miter saw like the one shown here cannot make compound miter cuts, but it does have advantages over a cheap plastic unit. Its saw is adjustable for tightness, and the blade can be changed. It quickly adjusts to make accurate 90- and 45-degree cuts and can be adjusted for odd angles as well. A series of holes with plastic "cams" clamp your work securely in place as you cut.

TIP **To be sure your saw can make the cuts you need, check its specifications (available online or at the store) to find exactly which size boards it can cut and at which angles.**

A sliding chopsaw uses a smaller blade but can make wide cuts and compound miter cuts. This one is set on a miter saw table, which holds it firm and provides support for boards on each side.

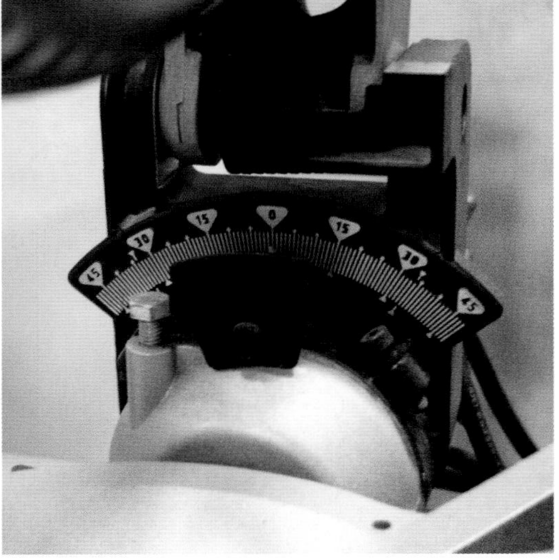

A double-compound miter saw tilts to each side, for quick cutting of compound miters in both directions.

Tablesaw

Though using a power miter saw is usually the easiest way to cut molding, a tablesaw comes in a close second, and it can also make long cuts that a miter saw cannot. If you're lucky enough to own a heavy-duty professional tablesaw—the kind that calmly hums when it's turned on—you've got a great tool for making all sorts of cuts. But even a mid-priced portable tablesaw reliably delivers accurate crosscuts, miter cuts, rip cuts, and even bevel and compound miter cuts. See pp. 61–63 for instructions on checking and calibrating a tablesaw to ensure precise cuts.

Most portable tablesaws nowadays quickly collapse for easy moving; some have wheels, whereas others can be simply carried by one or two persons. A 10-in. tablesaw is the most common size, and will easily handle all trimwork.

Here are some things to consider when choosing a tablesaw: The table should be large, so long pieces can rest firmly on its surface. "Outfeed support," usually in the form of a bar that can be pulled out, supports long pieces as you rip-cut them.

The rip fence should clamp firmly and be parallel to the blade. Some models, like the one shown below, have a removable fence, which you may need to check to be sure it is parallel to the blade; others (facing page) have a permanent fence that moves via a crank and stays parallel to the blade at all times.

If you will be cutting large plywood panels, you may want a fence that can be positioned 24 in. or more away from the blade. The saw should have a riving knife, anti-kickback panels, and a blade guard to protect you from material kickback and cuts.

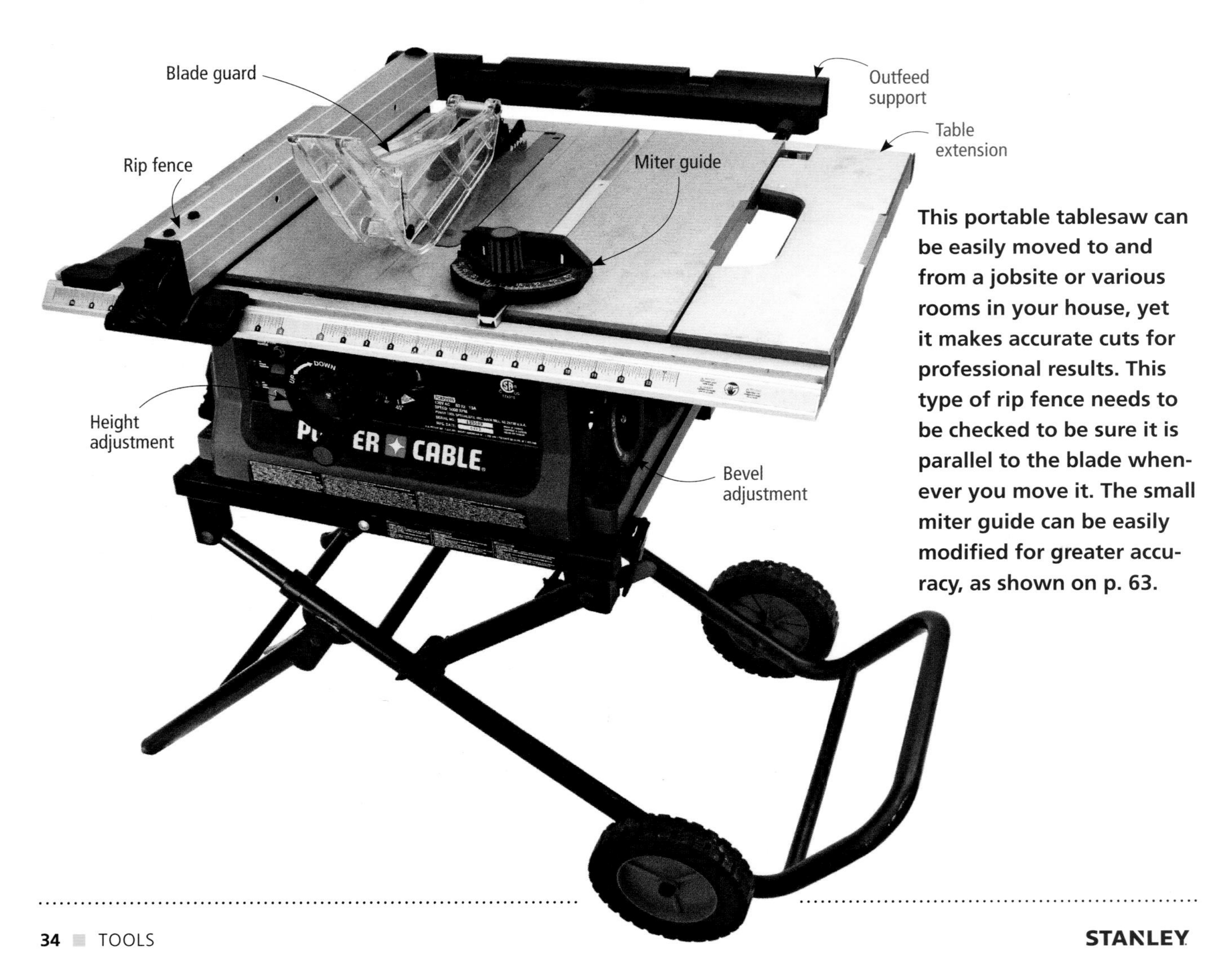

This portable tablesaw can be easily moved to and from a jobsite or various rooms in your house, yet it makes accurate cuts for professional results. This type of rip fence needs to be checked to be sure it is parallel to the blade whenever you move it. The small miter guide can be easily modified for greater accuracy, as shown on p. 63.

TIP A tablesaw can produce plenty of sawdust, so always use a dust collector (see p. 47) and wear a dust mask.

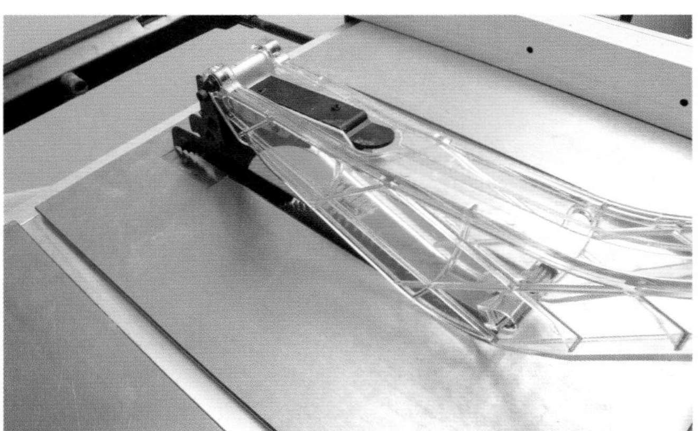

For the sake of clarity, most of the tablesaw photos in this book do not show this safety equipment. This setup has clear plastic guards to prevent touching the blades, a splitter that keeps the cut board separated as you cut, and anti-kickback pawls that keep the boards from lunging backward.

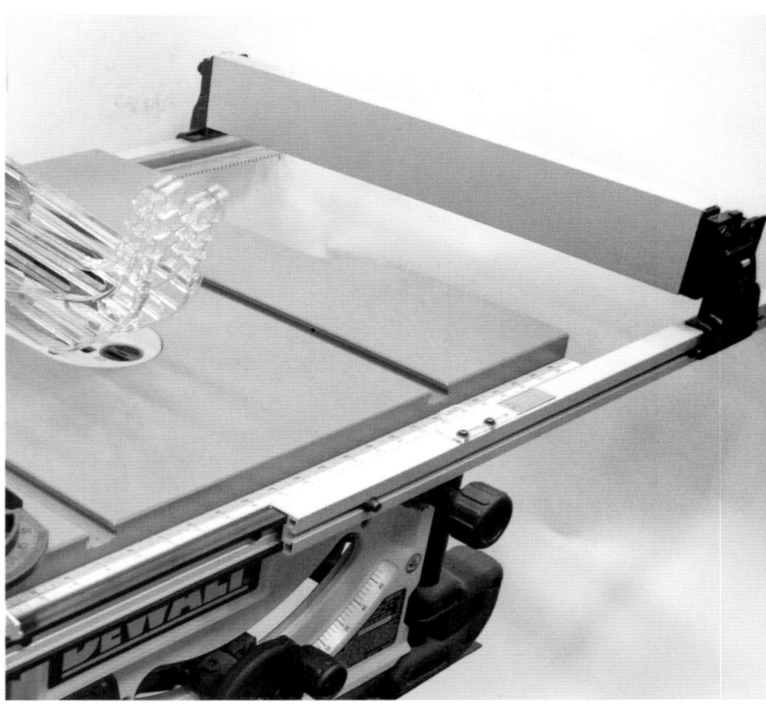

This tablesaw has a permanent adjustable fence, which stays parallel to the blade.

Radial-Arm Saw

Radial-arm saws are not as popular as they once were. But if you have one that's in good shape, it can make long, accurate crosscuts and miter cuts, and rip cuts as well. Set yours up with a dead-level table and a fence, as shown. Make test cuts and check that they are square or at 45 degrees. You may need to follow manufacturer's instructions for adjusting the saw for accuracy.

A lightweight circular saw can be used for general carpentry as well as some trim work. A blade guard shaped like this (right) is far less likely to bind when the saw is cutting thin boards or cutting near the end of a board.

Circular saw

A circular saw is not usually the tool of choice for cutting trim boards, but with care and the use of guides it can make very good straight crosscuts and rip cuts. A lightweight sidewinder-type saw, like the one shown at left, is more suited to trimwork than a worm-drive or a heavy saw.

A 7½-in. circular saw is the standard size. It will cut through 2× lumber even at a 45-degree angle. When choosing a circular saw, test the depth and angle adjustments to be sure they are easy to operate and firmly fix the base in place. The blade guard should be shaped so it will not bind on the wood, even when you are cutting thin boards or cutting near the edge of a board.

Some saws have an electric brake to stop blade rotation 2 to 3 seconds after releasing the trigger, for safety and to reduce the time you have to wait after making a cut. A pro-quality cord feels rubbery rather than stiff, and is less prone to damage. The base (or sole) should feel solid and perfectly flat.

For trimwork, use a fine-cutting blade with 60 to 90 teeth. Switch to a rough-cutting crosscutting blade when cutting framing.

TIP **An "orbital" jigsaw moves slightly forward and back as it moves up and down, for a slightly circular movement. This provides greater cutting power, so the saw does not need to work as hard and you do not need to push as hard.**

A cordless jigsaw with orbital action has plenty of power to cut smoothly through trim and 1× lumber.

The three most commonly used jigsaw blades are for general purpose (left), rough cutting (middle), and scrolling, for cutting tight curves (right).

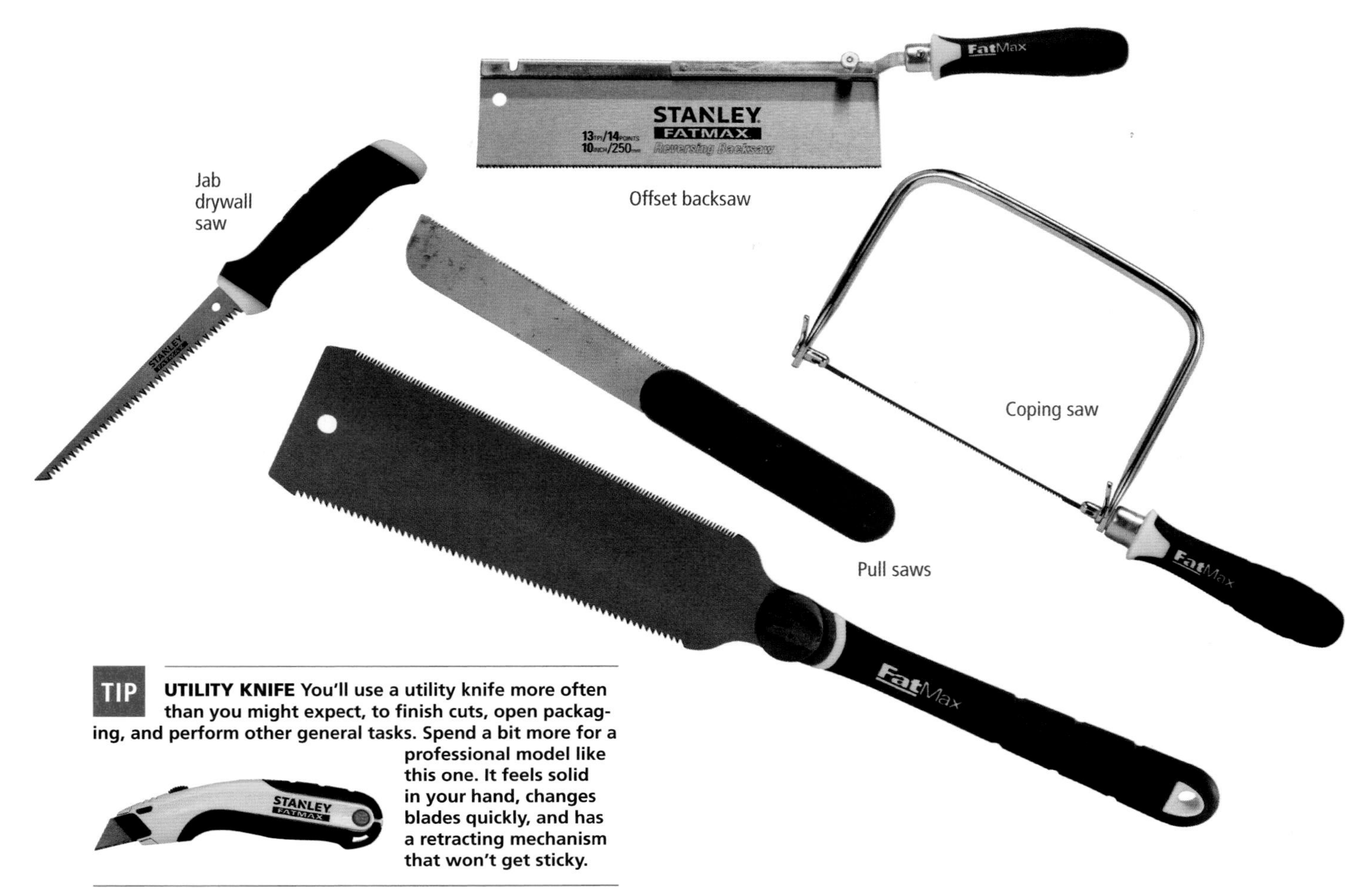

TIP **UTILITY KNIFE** **You'll use a utility knife more often than you might expect, to finish cuts, open packaging, and perform other general tasks. Spend a bit more for a professional model like this one. It feels solid in your hand, changes blades quickly, and has a retracting mechanism that won't get sticky.**

Jigsaw

A jigsaw (also called a sabersaw) makes curved cuts, and does so while creating a minimum of sawdust, which makes it useful for a variety of jobs. It is often used when making a cope cut (see p. 68). In a pinch, a high-quality jigsaw used with a guide can make straight cuts as well.

Older jigsaws had wobbly shoe plates and as a result blades were easily bent, making it difficult to keep a cut square to the surface of the board. But newer models are better built, so the blade will stay perpendicular to the shoe plate as you work. You can also adjust it for cutting 45-degree and other bevels.

Handsaws

Handsaws are still used in finish carpentry and are sometimes essential. These saws often give greater control for very fine cuts, and they produce very little sawdust. Handsaws can also be used to complete cuts made with a tablesaw or chopsaw.

You may need to use a jab drywall saw to cut away a section of drywall. Pull saws (sometimes called Japanese pull saws because of their country of origin) cut on the pull stroke, producing a nice straight cut. Some types are very thin, for fine cutting. A coping saw is used strictly for that purpose. A backsaw is rigid so it won't bend while you cut. The one shown above is offset, for cutting bottoms of casing to accommodate flooring that will slip under.

Renting Tools

If you have only a room or two to trim out and do not have a power miter saw or a set of nail guns, it may make economic sense to rent rather than buy either of these tools. Rented tools are usually of professional quality, and the rental salesperson can quickly show you how to operate them. When renting nail guns, you will need to purchase the nails.

Demolition can be sweaty and gritty, so make it easy on yourself and keep a set of prying tools handy.

Demo Tools

Often old molding needs to be removed before the new can be installed, and sometimes you will need to correct mistakes. Tools for prying boards and removing nails are essential for these unglamorous tasks. Whenever possible, use tools that do not damage the wall nearby while you work. A pair of taping knives or straight scrapers will get you started. A flat prybar is the most useful tool for prying trim. To pull out nails when you don't need to save the trim board, use a cat's paw. For more surgical removal, after the board has been pried off the wall, use a pair of locking pliers or nippers.

TIP **Different tape measures sometimes produce slightly different measurements, so it's usually best to stick with one tape measure on the job.**

The lower tape measure is 1¼ in. wide, with standard markings. The upper one is 1 in. wide, and has markings indicating eighths of inches. Its red markings indicate a measurement from the back side of the tape body, for when you are measuring between walls or other surfaces.

Measuring Tools

Though the best way to measure a board for cutting is to hold it in place and mark, you will often need to use measuring tools. Choose tools with numbers that you can read easily.

Tape measure

A tape measure is a trim carpenter's constant companion. A 25-ft.-long tape suits most purposes, but if you will work in large rooms, a 30- or 35-footer may be a better choice.

A tape measure's hook, attached to the end of the blade, is slightly loose, so it can move back and forth by the same distance as the thickness of its metal—which means you will get an accurate measurement whether you are hooking it to the end of a board or pressing it against a wall or other surface.

Choose a tape measure that you find easy to read and easy to handle. A high-quality tape measure will have a special coating to resist smudging and keep it easy to read. It should slide smoothly out and in. While measuring, you sometimes need to extend the tape outward, so choose a model that can extend at least 8 ft. without bending. A 1¼-in.-wide tape measure often can extend to 11 ft.; a 1-in.-wide model may extend only 8 ft.

Squares

Trim often must be cut and installed square, and simple tools help keep things straight. A small Speed Square® is the most commonly used because it's quick to align and very durable. It makes for easy marking and checking of 90- and 45-degree angles; the other angles printed on the tool are not very accurate and should not be relied on when doing trimwork. In addition, a very large (and sometimes orange) Speed square, as well as a framing square, help check layouts for square.

Many carpenters love their combination square, which has a blade that easily slides into the most convenient position. It checks for 90 and 45 degrees even in tight spots, and helps measure or scribe straight lines for reveals (see p. 124). It also has a level bubble, so you can mark level and plumb lines. And it's also a small ruler, handy for making quick short measurements.

A sliding bevel, also called a T-bevel, can capture odd angles so you can reproduce them on boards for cutting.

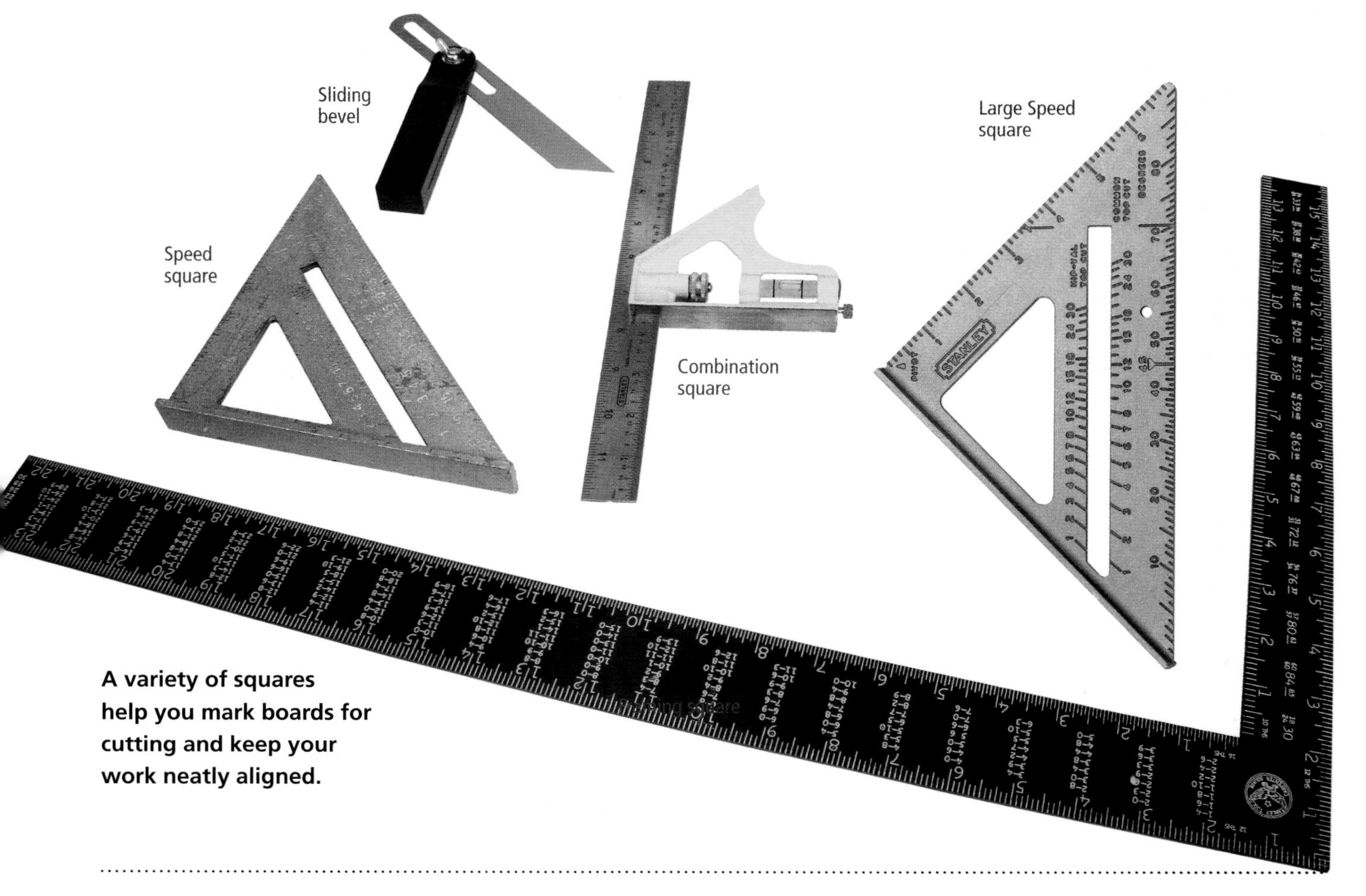

A variety of squares help you mark boards for cutting and keep your work neatly aligned.

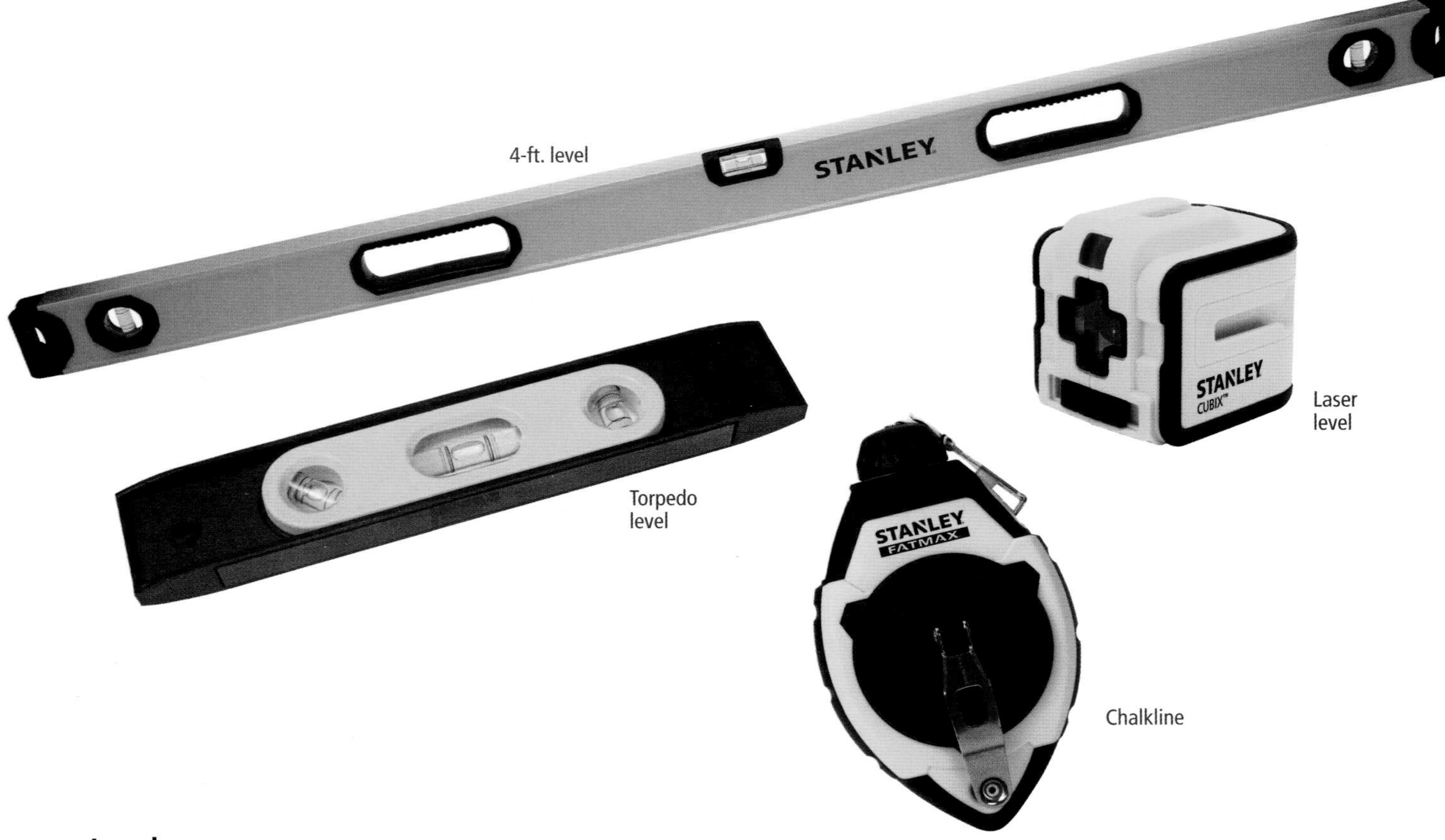

Levels

Baseboard, crown, and other moldings simply follow the line of the floor or ceiling, but other moldings, such as chair rail and paneling, need to be installed level and plumb. And doors should always be checked for plumb and level as well.

Bubble levels, sometimes called spirit levels, have been keeping things straight for centuries. A short "torpedo" level is useful in tight spots, and a 2-ft., 4-ft., or 8-ft. level will help you check everything else.

Newer laser levels project level and plumb beams of light onto a wall, so you have a continual frame of reference and don't have to constantly stop to check your work. To mark a long, straight line, use a chalkline.

Levels check that your work is level (perfectly horizontal) or plumb (perfectly vertical), so a trimmed-out room feels comfortably in harmony with the world.

Tools for Shaping and Smoothing

Trim boards and wall panels sometimes need to be modified after cutting and before installation. Tools that plane, rasp, chip, and sand are all helpful from time to time.

Planes shave boards smooth and can also straighten out uneven conditions. A block plane is the most common type used for trimwork. A small trimming plane can straighten uneven miter cuts. Surform® tools remove material quickly, but leave a rough surface behind, which you can sand or plane smooth.

A very sharp chisel can be tapped with a hammer or pushed by hand to pare away material and clean up cuts. Chisels often finish or make cuts in places where no other tool can reach.

Trim boards and panels often have rough spots that need to be smoothed. Power sanders can make quick work of this. Use a belt sander—and use it carefully—only if you actually want to change the shape of a board, because it can eat away with surprising speed. A detail sander or random-orbit sander moves more gently, but still gets the job done. For final finishing, use hand sanders.

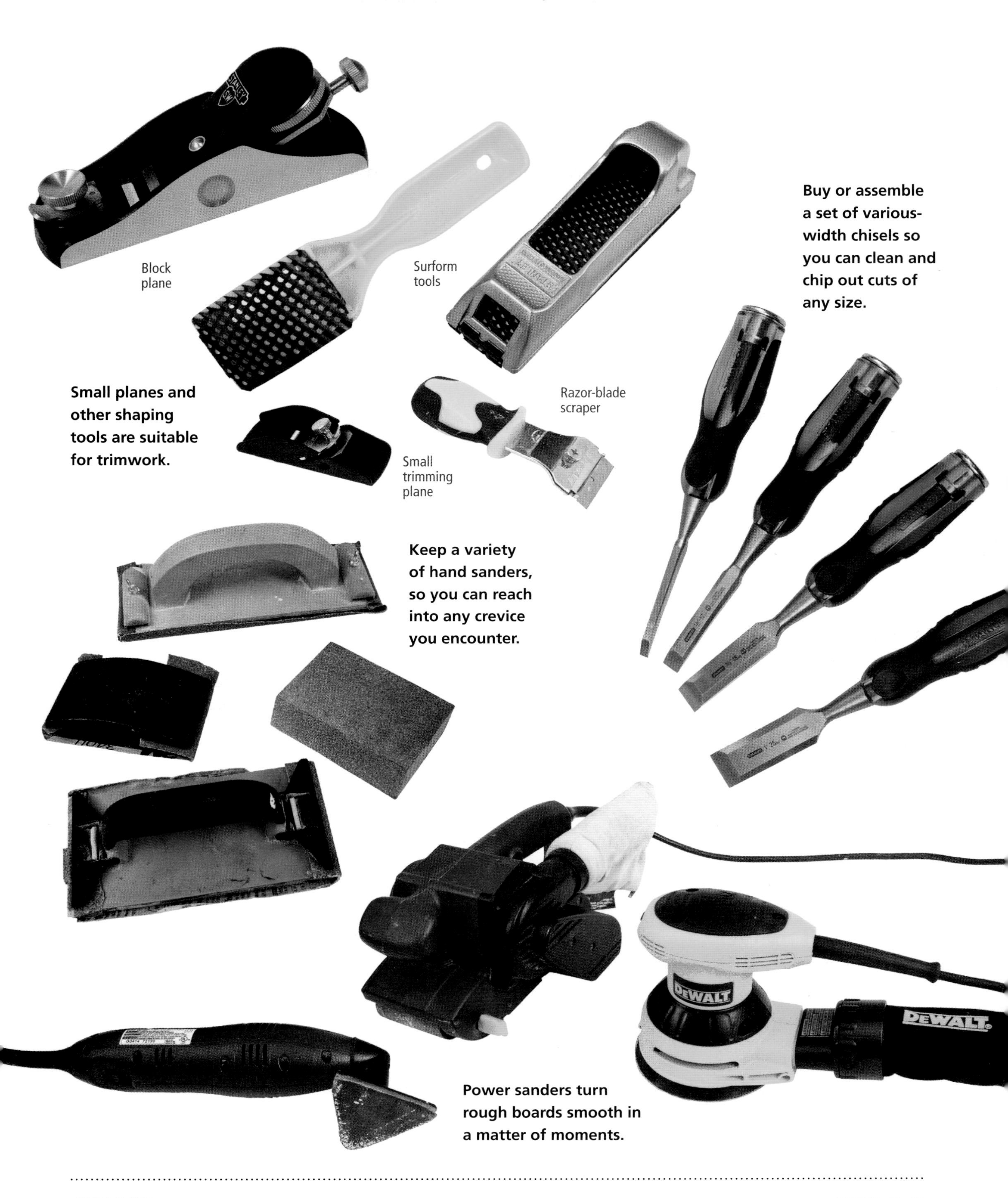

Small planes and other shaping tools are suitable for trimwork.

Buy or assemble a set of various-width chisels so you can clean and chip out cuts of any size.

Keep a variety of hand sanders, so you can reach into any crevice you encounter.

Power sanders turn rough boards smooth in a matter of moments.

Router

A router can be used to ease a too-sharp edge, and can also create molding out of plain boards. See pp. 76–77 for examples of making your own trim using a router. Various router bits can create just about any profile you can imagine.

A router for trimwork does not have to be particularly powerful, and it doesn't need fancy features. A fixed-base 1¼-horsepower router will work just fine. If you want to do elaborate scrollwork or work with bits that are wider than 1 in., then get a 2-horsepower router. Understand how to adjust the router for depth, and see that you can make microadjustments without difficulty.

TIP **Make sure your router can handle the bits you want to use—either ¼ in. or ½ in. Some routers can be adapted so that bits of either size will fit.**

1 This router has an adapter so it can be fitted with either ¼-in. or ½-in. router bits. 2 Choose a router that can be easily adjusted for depth.

Biscuit Joiner

A biscuit joiner, also called a plate joiner, attaches pieces by incising slots in both pieces, which accept "biscuits" made of pressed wood. Cut slots in each piece that align with each other, squirt in some glue, slip the biscuits into one of the pieces, and press the pieces together. Clamp the work tightly for at least an hour, and you have a very tight joint. Biscuits are especially useful for assembling panels (see pp. 74–75).

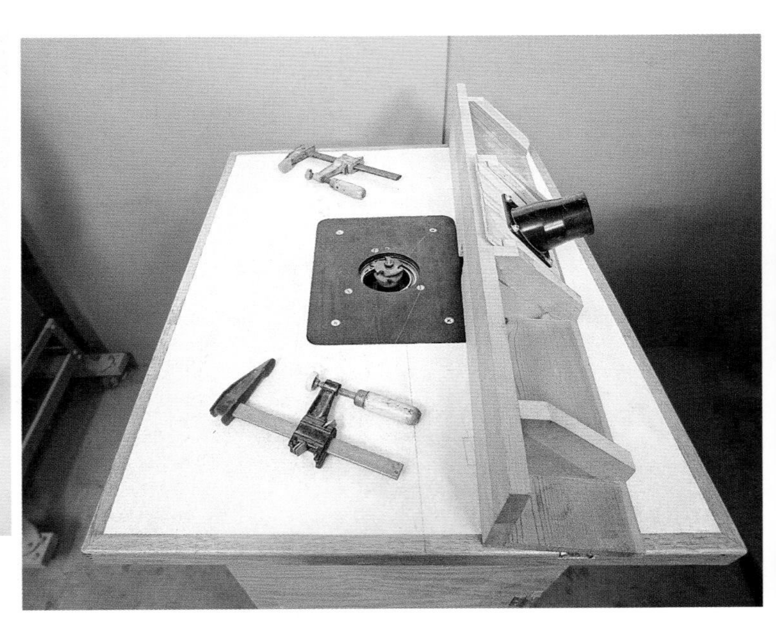

You can buy a router table or make one yourself. A table makes it possible to create trim pieces out of straight lumber, or to modify trim pieces by adding additional profiles.

Fastening Tools

A set of nail guns may set you back a few hundred dollars, which is a good deal if it saves hiring a pro for a couple of days. For small jobs, you can get by with hand nailing.

Nailers

Also called air guns, pneumatic nailers are definitely the tool of choice for people who do a good deal of trim carpentry. The advantages are many: Using a nailer, you can drive nails with one hand while you hold the board with another. The nails are thinner and have smaller heads, so they are easier to fill and sand—or often, they don't need filling at all. And although it's easy to miss your mark with a hammer and dent a board, that will not happen with a nailer.

A modestly priced set may include two or three nailers and a pancake-style compresser. This setup may not last long if you are a professional who works with it daily, but it can be a good old friend for a homeowner.

Gas-powered and cordless battery-powered finish nailers are also available. These generally cost more than pneumatic nailers but can allow you to move around more easily without being tied to an air compressor and air hose. Cordless nailers can be a good value if you already have other cordless power tools that use the same batteries.

Hammer

Even if you attach all your trim with nail guns, you will still need a hammer from time to time—for instance, for demolition and for tapping boards into place.

TIP **A typical air gun set includes a framing nailer and a stapler, which you may not need often for trimwork. You will use one or two finishing nailers, and you may need to buy a separate pin nailer.**

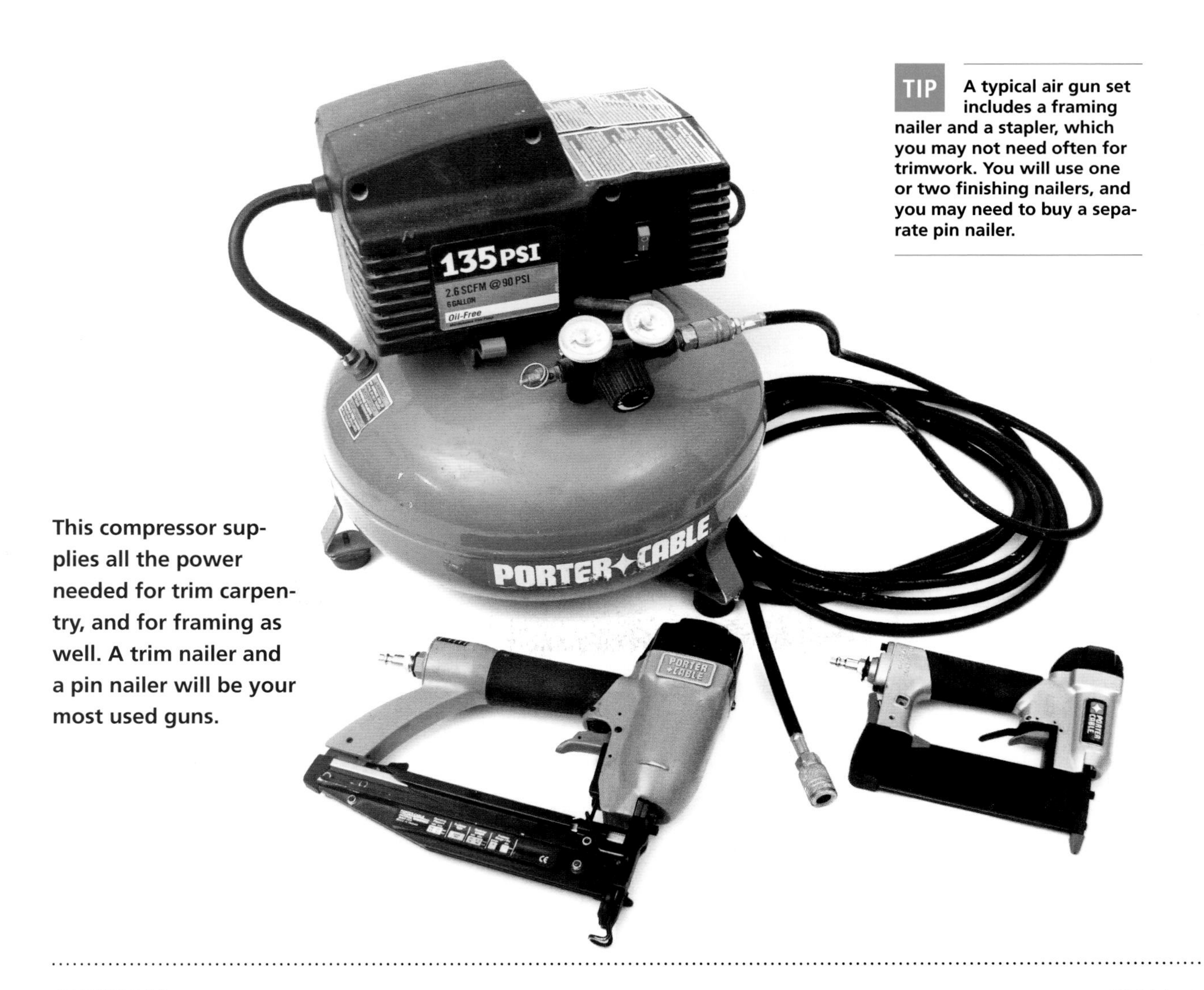

This compressor supplies all the power needed for trim carpentry, and for framing as well. A trim nailer and a pin nailer will be your most used guns.

TIP Some people prefer a hammer with a curved claw because it gives good leverage when removing nails. Others prefer a straight claw because it is more versatile and better for prying boards. It's your choice.

This hammer is ergonomically designed to be easy on your joints while you pound.

Countersink bits drill a pilot hole, a countersink hole (a shallow hole shaped like a screw's head), and a counterbore hole as well, if you keep pushing. This allows you to drive standard screws with their heads either flush with the surface or sunk below the surface; you can fill the resulting hole with a dowel or with wood filler.

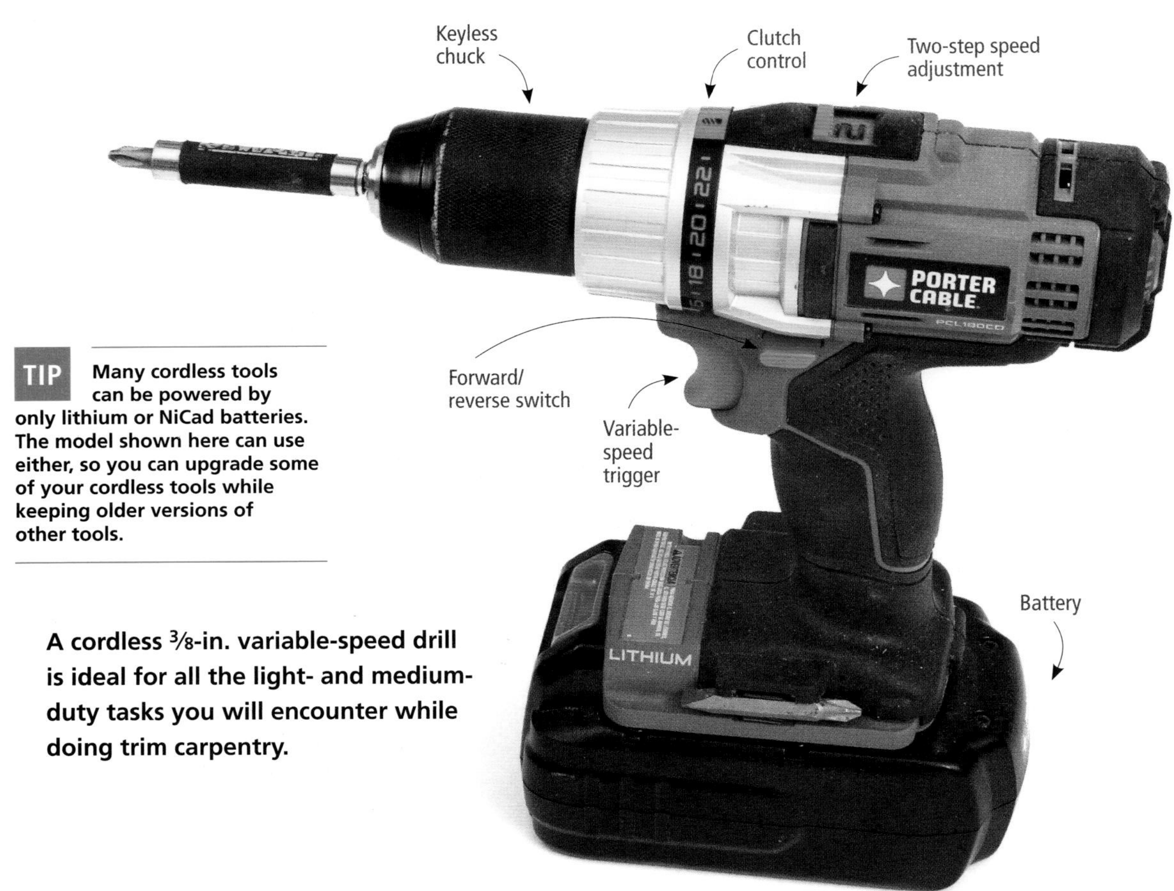

TIP Many cordless tools can be powered by only lithium or NiCad batteries. The model shown here can use either, so you can upgrade some of your cordless tools while keeping older versions of other tools.

A cordless ⅜-in. variable-speed drill is ideal for all the light- and medium-duty tasks you will encounter while doing trim carpentry.

Select a hammer that feels comfortable in your hand as you swing it. A wood-handled hammer absorbs shock fairly well and is a good inexpensive choice, or you may want to upgrade to a modern anti-vibration model. A 16-oz. model is ideal for attaching trim. There is no need for a heavier hammer, and you may want to go with an even lighter one—say, 12 oz.

Drill

Use a drill (often called a drill driver) to bore pilot holes before driving fasteners, and use it to drive trimhead and other screws. Today most people use cordless drills; one with an 18-volt or 20-volt battery will take care of all your drilling and driving needs.

A ⅜-in. drill is the right size. (You need a ½-in. drill only for heavy-duty tasks.) A good drill has a light that shines on the work—a real help in dim conditions. Check out the battery's "amp hour" and "charge time" ratings (you may need to go to a website and look under "specifications"). The battery should take no more than 1 hour to charge and should last at least 1½ hours while in use. Be sure to have an extra battery so you can charge one while you use the tool.

A clutch control is helpful, especially when attaching softwood boards; it controls the torque, so you don't overtighten and perhaps drive the head all the way through the board. A two-step speed adjustment is also a good feature, because you sometimes want to drive speedily and other times want to drive carefully. The keyless chuck should be gnarled or ridged in a comfortable way so you can hold it firm while turning on the drill to change bits.

Pocket Screws

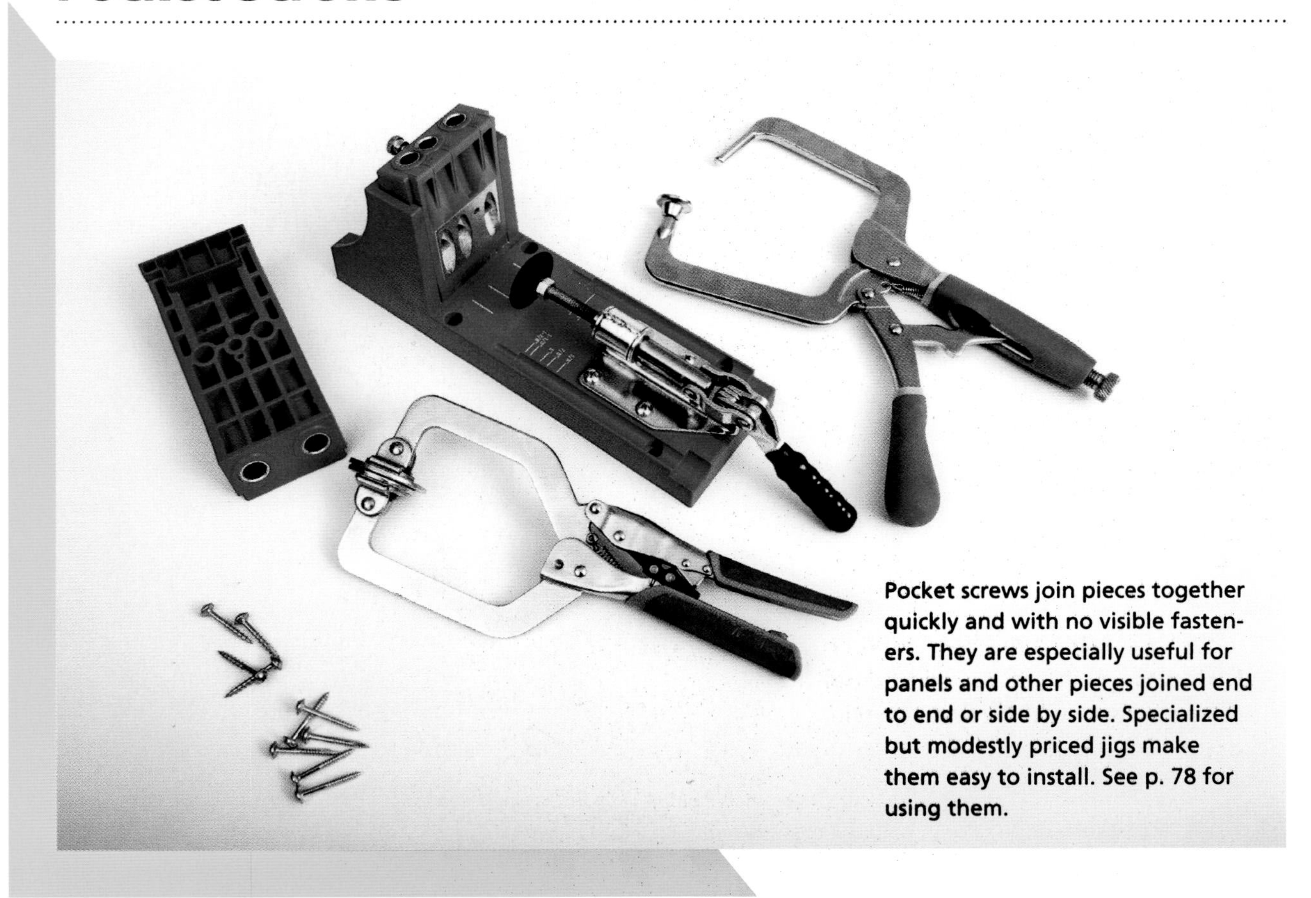

Pocket screws join pieces together quickly and with no visible fasteners. They are especially useful for panels and other pieces joined end to end or side by side. Specialized but modestly priced jigs make them easy to install. See p. 78 for using them.

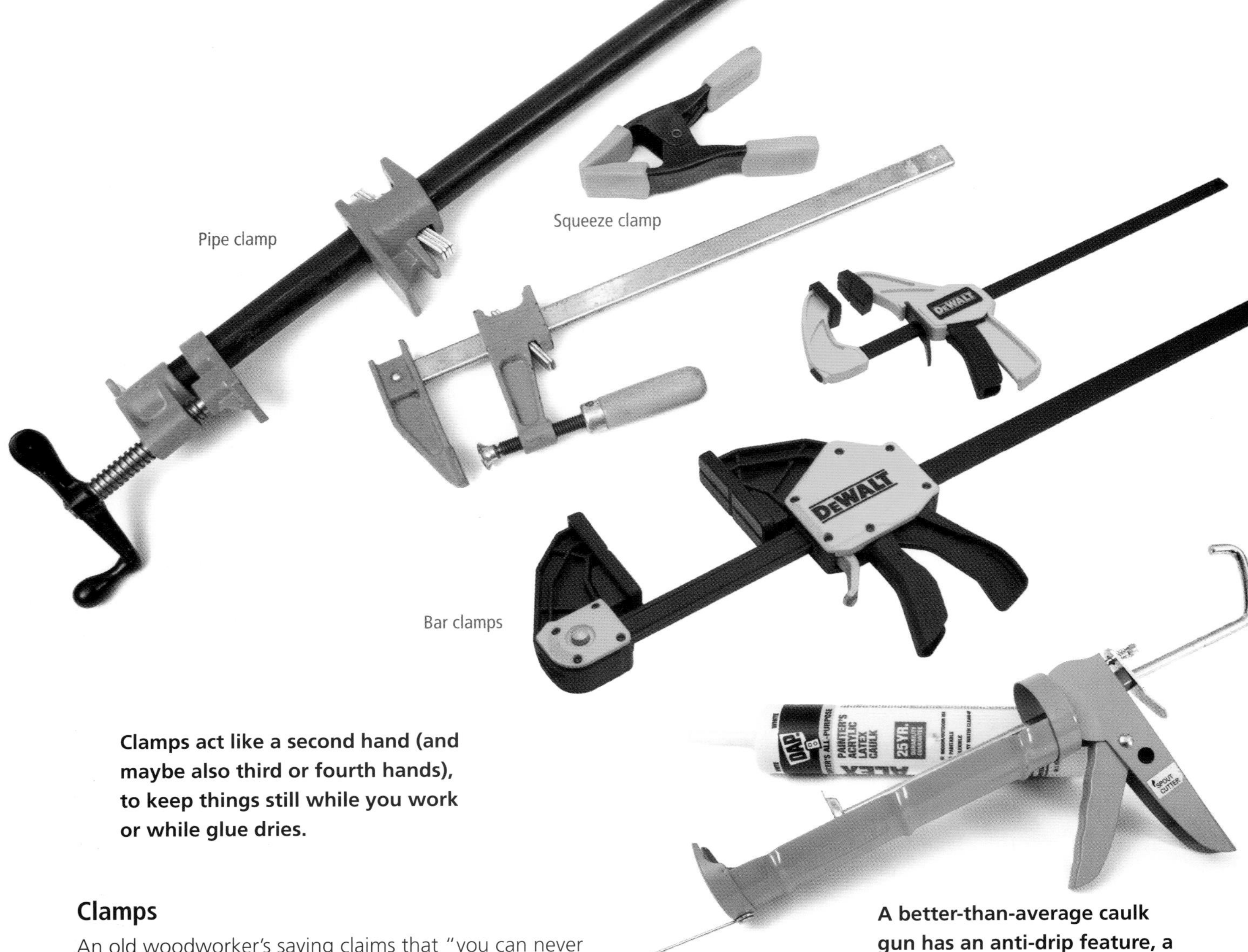

Clamps act like a second hand (and maybe also third or fourth hands), to keep things still while you work or while glue dries.

A better-than-average caulk gun has an anti-drip feature, a seal punch, and a spout cutter.

Clamps

An old woodworker's saying claims that "you can never have enough clamps." That is especially true of cabinetry, but clamps are also often useful for trimwork. Use clamps to hold work in place while you sand or shape it, and to squeeze glued-together parts tight while the glue sets.

Situations vary, so variety is good. Small squeeze or pinch clamps are quick to attach and hold with modest power. Bar clamps take only a bit more time to attach, and they hold as firmly as you choose to turn the screw or squeeze the trigger. For large work, use pipe clamps, which are as long as the pipe you attach them to.

Caulk gun

If you will paint your trim, you'll probably caulk some of the joints, including where trim meets a wall. Buy a good caulk gun, and keep it as clean as possible.

Many caulk guns will keep pushing caulk out of the tube after you've stopped squeezing the trigger, producing a mess and making the job difficult. Better caulk guns have a "dripless" or "anti-drip" feature, which allows the push rod to back up slightly when you release the trigger and greatly reduces the annoying ooze.

Some caulk tubes have a seal that needs to be poked through before the caulk will flow, so choose a gun with a seal punch. A spout cutter is also a handy feature, though you may end up cutting spouts with a knife anyway.

Dust Collection

A chopsaw, tablesaw, or circular saw will throw out clouds of dust. That's fine if you're cutting outdoors, and maybe OK if you're working in a shop, but on most jobsites dust can be a real problem, not just for keeping things clean but also for your health. Nowadays there are a good number of dust collection options, with wide variations in size, price, and effectiveness.

The simplest solution is to hook up a shop vacuum to the saw's dust port. The effectiveness of this approach depends on the vacuum. The modestly priced vacuum shown below has a HEPA filter, which will do a good job of keeping dust from leaking through the machine and back out into the room. The motor is medium strength, so you can probably count on this setup to suck away 80 to 90 percent of the dust.

A stronger and more elaborate dust collector like the one shown below right will do a more thorough job. But a large unit like this is better suited to a shop than a jobsite.

Some saws are more amenable to dust collection than others. Most circular saws do not have dust collection ports, so all you can do is have a helper hold a vacuum's tip near the blade as you cut, which may reduce dust by 60 percent or so. Many chopsaws and tablesaws of recent manufacture are designed for efficient dust removal, but many older models throw out a good deal of dust even if you hook up a collector. Some tablesaws are designed so that most of the dust gets sucked into the collector and almost all the rest falls down onto the floor below rather than getting blown into the air.

TIP **To connect your dust collector to a saw, you may be able to find a factory-made connector that seals tightly and is easy to attach and remove. Or you may need to cobble together your own. Hint: Connectors like rubber hose clamps are far more reliable than any arrangement that involves duct tape.**

A shop vacuum has a less powerful motor than a dust collector, so the hose cannot be long. One with a HEPA filter will keep the air pretty clean.

A shop-type dust collector has a powerful motor, so hoses can extend 20 ft. or so and still suck effectively. You may be able to connect it to several tools, so you don't need to keep changing hoses every time you switch tools.

CHAPTER FOUR

TECHNIQUES YOU'LL USE OFTEN

THROUGHOUT THIS BOOK you'll find specific instructions for cutting and installing different types of trim. This chapter shows some of the most basic and common methods that are used over and over again in the course of trimwork.

Start with a good collection of tools, as shown in the previous chapter. You don't have to buy them all at once, but when you encounter a task that could be made easier with an inexpensive new tool, don't hesitate to buy it. You might get by without a specialized tool, but the additional struggle that comes from using a wrong tool can also lead to unattractive mistakes that you will regret every time you look at the trim.

Don't hurry. Working hastily often leads to sloppy results, and it will take you far more time to correct the resulting mistakes than it would have taken to do it right the first time. Measure twice, and check measurements by holding boards in place whenever possible. Make a flat and easy-to-use work surface that you can lay boards on when cutting. And every once in a while stand back to examine your work, so you can spot things that should be changed sooner rather than later.

Removing Old Trim

If you're replacing old trim with new, the first step is to pull off the existing trim, with the goal of doing as little damage as possible to the wall. If the old trim is stained, it may come off easily. There should be no connection between wood and wall, and nail heads should be findable (just under the spots of wood filler). But painted trim is typically joined to the wall with paint and often caulk as well, and the nail heads will be hard to find.

TIP **You may be tempted to remove the old trim, strip and stain it, and reattach. Do this only if you are sure the wood is in good shape. All too often, older painted trim has cracks and numerous nail holes. Stripping the wood can be very time-consuming, especially if the trim is elaborate. The best option may be to take the trim to a professional stripper, who can dunk the boards in vats of stripping solution. This will be expensive, but may be worth it if the trim is made of expensive hardwood.**

1 **CUT THE PAINT. Use a utility knife to cut through the paint and perhaps also the caulk where the trim board meets the wall (top). Cut all the way through, until you feel a dead space. If you don't do this, the paint will chip on the wall, leaving you with a time-consuming patching and painting task. Also cut the line where the trim meets other trim—in the example above, where the door casing meets the jamb.**

2 **PUNCH NAILS. If you can see the nail heads, or a slight indentation showing the presence of a nail, you may choose to punch the nails all the way through the trim board, using a hammer and thin nail punch. Once fully punched, they will not hold the trim.**

3 **USE TWO TAPING KNIVES.** Tap a taping knife or straight paint scraper into one of the lines you cut in Step 1, near an end of a trim board. Then tap in another on top of it. This will pull the trim away slightly, without damaging the wall.

4 **PRY THE TRIM LOOSE.** Switch to a flat prybar to remove the trim. Work from one end toward the other, and position a taping knife or a scrap piece of wood under the bar to protect the wall. You may need to pry both sides of the trim in this way.

5 **PULL NAILS.** Once the board is removed, use a pair of slip-joint pliers or nippers to remove any nails in the trim board and on the wall. Grab the head of the nail with the end of the pliers' jaws, squeeze hard, and roll the pliers back to ease the nail out.

Checking Openings, Doors, and Windows

Before you install trim around a door or window or against abutting walls and ceilings, it helps to check to see that they are square, as well as plumb or level. If corners are out of square, you may have to cut trim pieces to odd angles, which can be difficult and can take extra time. In some cases you can adjust the door or window—or even a surface—so that trim pieces cut at good old 90- and 45-degree angles will fit just fine.

TIP **A window or an exterior door will be attached to outside trim as well as inside trim. To straighten the jamb out, you may need to remove the exterior trim, or at least cut through some of the nails that fasten the exterior trim to the jamb.**

CHECK WITH SPEED SQUARE. For a small opening, you can check with a small metal or a larger plastic Speed square. Press the square into the corner; it should fit snugly along both sides. Check both upper corners of a door and at least two corners of a window.

CHECK WITH FRAMING SQUARE. For a larger opening, use a framing square, which is a bit more accurate because it has longer legs.

CHECK DIAGONALS. You can also check by measuring diagonals. Measure from corner to corner. Be sure to hold the tape's hook in the same way for each measurement. If the measurements are the same, then the opening is square.

Quick Fixes

In some cases you can quickly straighten out an opening. If a stud has globs of dried joint compound, scrape them off with a paint scraper or the side of a hammer's head.

1. Remove or pound in any protruding nails or screws.
2. In some cases, you can solve a slightly out-of-square stud by pounding on it with a hammer.
3. Finally, make sure drywall does not protrude into the opening. If it does, cut it back with a knife or a handsaw.

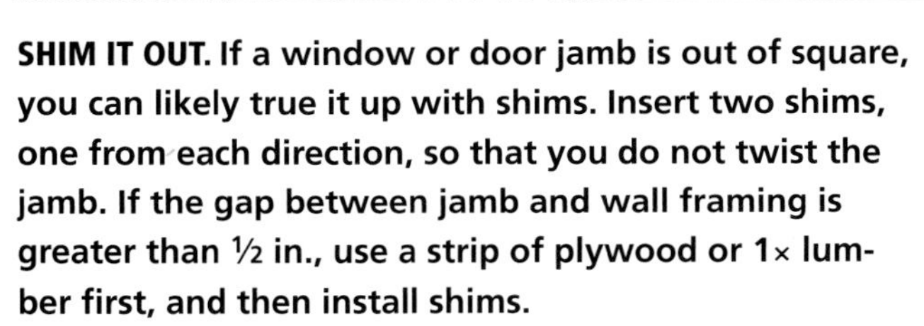

SHIM IT OUT. If a window or door jamb is out of square, you can likely true it up with shims. Insert two shims, one from each direction, so that you do not twist the jamb. If the gap between jamb and wall framing is greater than ½ in., use a strip of plywood or 1× lumber first, and then install shims.

FIX WOBBLE. Use a long level or other straight edge to check framing members for straightness. If the level wobbles, or if there is a gap in the middle, then shim the jamb carefully to achieve a straight line.

Measuring and Marking

Develop good work habits when measuring and marking and work in the same way every time. Whenever possible, hold a board in place, with one end pressed against a wall it will abut, and mark the other end for the cut. But often, measuring with a tape measure is the only possibility.

TIP **Become accustomed to the markings on your tape measure, so you immediately recognize fractions like 3⁄8, 3⁄4, and 5⁄8. If a measurement falls on one of the tiny lines indicating sixteenths—say, 11⁄16 —it may be too complicated to figure out. Then it's easier to start with the nearest quarter-inch and describe the measurement as a sixteenth "minus" or "plus." So, for instance, call 5 11⁄16 "five and three-quarters minus," and call 8 5⁄16"eight and one-quarter plus." (Write these measurements as "5 3⁄4-" and "8 1⁄4+.")**

CHECK END FOR SQUARE. Before measuring, check that the end you will not cut is square. If it is not, cut it.

MARK IN PLACE. Hold a board in place and mark with a pencil or knife for cutting to length. Sometimes it is best to mark the back side of the board. Hook the tape onto one end of a board, pull it tight, and mark the location of the cut. Some carpenters use a simple short straight line, whereas others draw a V, with the point of the V being the exact place to cut.

DRAW A SQUARE LINE. If you are cutting with a chopsaw, this step may not be needed (see p. 57). To draw a square line across the board, place the tip of the pencil on the cutline, slide the square over to touch the pencil, and then draw the line. Draw an X on the waste side of the cut, so you will know to position the sawblade on that side of the line. (A sawblade is often 1⁄8 in. thick, and if you cut on the wrong side of the line the board will be 1⁄8 in. short.)

Marking for a Miter

1. To measure for a miter cut, first measure and make a short line to mark either the heel or the tip of the cut.

2. Hold the pencil on that spot and slide a square over to touch the pencil, then draw the angled line.

3. Marking for a miter on a board with Colonial or other details can be tricky, because the pencil will wander as it encounters hills and valleys. Draw a line as shown, but be aware that the line will be more accurate at the high points; the low points may not be accurate.

BURN AN INCH. When you cannot hook a tape's hook to the end of a board—as when measuring from line to line or between inside edges—you'll get a more accurate measurement if you "burn an inch": Hold the 1-in. mark against the line or board at one end as you measure. Of course, be sure to subtract that inch when you cut the board.

Cutting with a Miter Saw

Though you could use a tablesaw or a hand miter saw, many pros find a power miter saw the best tool for cutting all kinds of trim. Equip it with a fine-cutting blade that has at least 60 teeth, and provide a table or supports to hold long boards level with the saw's table while you work.

Setting up

Before you start cutting boards, check to be sure that the saw is cutting accurately, and provide an easy way to support boards.

To test for cutting accuracy, set and lock the saw angle at 90 degrees and make a cut. Flip one of the cut pieces over and push it against the other piece while holding both boards firmly against the fence or another straight edge (below left). The two pieces should match perfectly. If not, the angle is off; follow manufacturer's instructions to micro-adjust the angle. Also make a 45-degree cut and position the cut piece as shown (below center). The two boards should form a perfect 90-degree angle.

For a very simple support system, make moveable blocks. Cut two pieces of ¾-in. plywood about 12 in. long and as wide as the height of the saw's table. Cut another piece to the same length and ¾ in. narrower than the others. Assemble as shown (below right). This block can be used simply to support pieces, or, if you clamp or screw it to a workpiece as a length guide, for cutting multiple pieces to the same length (top photos facing page).

If you have the luxury of a shop, screw the miter saw to a workbench and set up a long working surface on each side, at the same height as the saw's table. You can buy a measuring track with an adjustable stop, as shown in the bottom left photo on the facing page. Once it is precisely adjusted, you can position the stop at the correct measurement, press the board against it, and make the cut, saving plenty of measuring and marking time.

Cutting tips

A great advantage of a miter saw is that it nearly always makes clean cuts, with no visible tearout (a series of small splinters along a cutline). In most cases you will cut with the board's face up, so any small splinters and tearout will be on the underside, which won't show once the board is installed. As the photos here show, even a "general purpose" blade, if sharp, can make crisp cuts.

CHECK 90-DEGREE CUT. Make a cut at 90 degrees, flip the cut piece, and butt against the other piece. The two pieces should match.

CHECK 45-DEGREE CUT. Make a cut at 45 degrees and position the cut piece between the other piece and a square.

MAKE A SUPPORT. Assemble three pieces of ¾-in. plywood to make a support block.

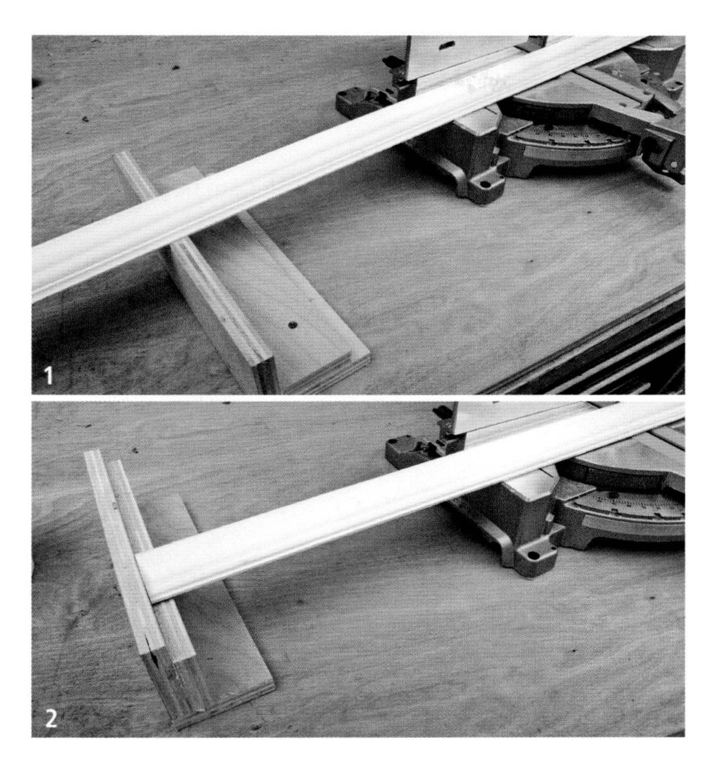

SUPPORT OR STOP. Use as a support for long boards (1) or as a stop guide (2).

Miter-Saw Safety

A miter saw's guard retracts to fully expose the blade during a cut. Keep fingers, hands, and body away from the blade, outside of the hazard zone marked on the base of the saw.

- As is the case when operating any power saw, wear short sleeves, or roll back long sleeves, so there is no chance that clothing can catch in the saw. Do not wear gloves.
- Wear safety glasses. If you are working in a small room, also consider ear protection.
- If possible, attach a dust collector or shop vacuum to the saw's dust collection port. Otherwise, it's a good idea to wear a dust mask.
- Clamp or hold the board tightly against the fence and base, keeping your hand outside of the hazard zone.

SHOP SETUP. Secure the miter saw to a workbench and set up a working surface on either side.

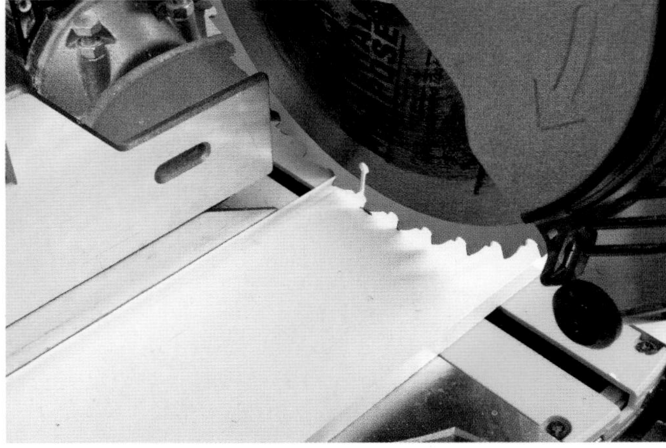

LINE UP THE BLADE WITH A SHORT CUTLINE. Though you can draw a cutline with a square, as shown on p. 54, you can save time by simply drawing a short cutline near where the blade will meet the board. With the saw not running, lower the blade until a blade tip meets the line, and move the board into precise position. Raise the blade, turn on the saw, and make the cut.

Miter cuts

A chopsaw excels at making reliably accurate 90- and 45-degree miter cuts. Where the boards need to cover a corner that is not exactly square, you can make small adjustments.

Micro-adjusting

Sometimes you may need to make cuts that are slightly greater or lesser than 90 or 45 degrees. (The two most common examples are when a door jamb is out of square and when a wall corner is out of plumb.) If your miter saw has a reliable miter guide with a firm angle lock, you can set the angle to, say, 44½ or 90¾ degrees. Another time-honored solution is to use very thin shims, such as playing cards (see the photos on the facing page).

TIP **Many chopsaws move suddenly down and up, in a way that is difficult to control, at the moment you pull the trigger as well as when you release the trigger. To counter this, keep the blade well away from the board until the saw is running and you have full control; then make the cut. And lift the blade away from the cut before releasing the trigger.**

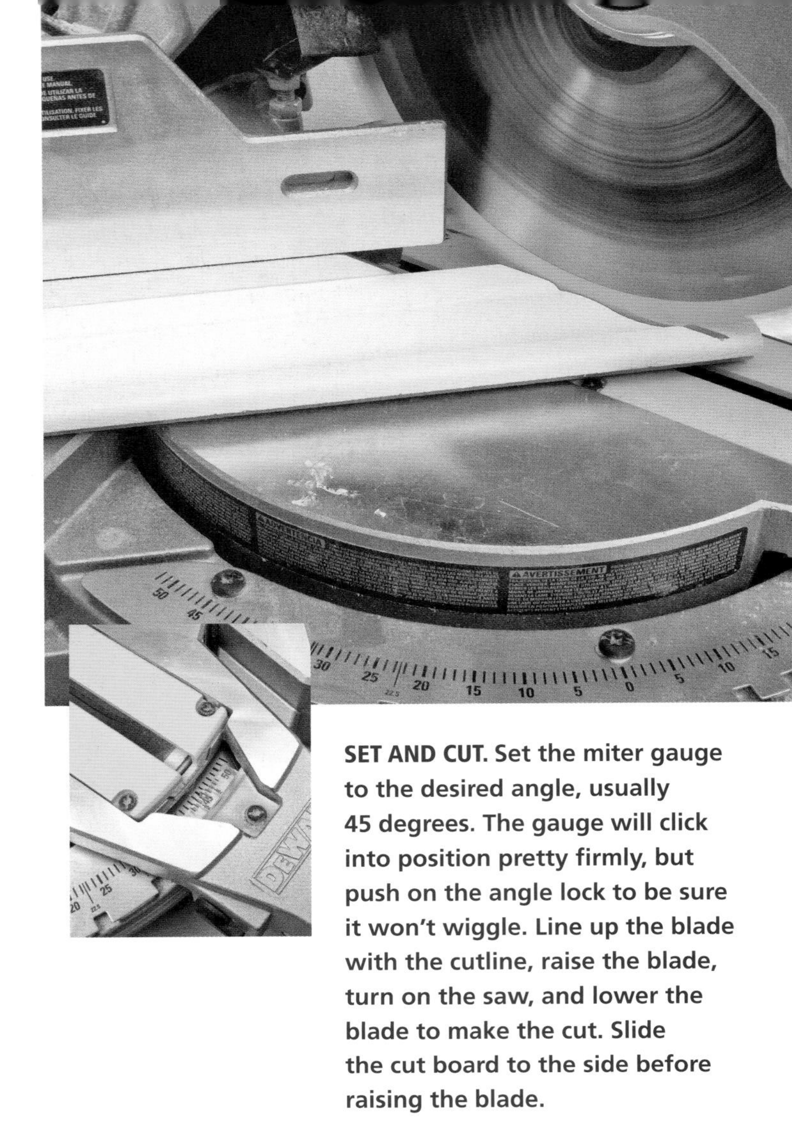

SET AND CUT. Set the miter gauge to the desired angle, usually 45 degrees. The gauge will click into position pretty firmly, but push on the angle lock to be sure it won't wiggle. Line up the blade with the cutline, raise the blade, turn on the saw, and lower the blade to make the cut. Slide the cut board to the side before raising the blade.

Avoiding Tearout

If you raise a running blade against a just-cut board, or if you release the trigger while the blade is lowered, you could end up with tearout. So lower the blade and cut completely through the piece, then release the trigger and let the blade come to a complete stop before raising the blade.

MICRO-ADJUST WITH CARDS. On a 45-degree miter cut, place a shim (or two or three) near the blade (1) or away from the blade (2), depending on how you want to modify the angle.

TIP It can be difficult to draw an accurate miter cutline on a piece of trim with a wavy surface. To be safe, many pros make a first cut that's about ⅛ in. too long. Then they "sneak up" on the cutline, shaving away small increments until they reach the exact point.

Two Ways to Bevel-Cut

Make bevel cuts on scrap pieces and test to be sure that (1) the bevel is the correct angle and (2) the cutline is at an accurate right angle. If your saw is not a compound miter type, there is only one way to make a bevel cut: with the board held upright against the fence (**1**). This is usually the most reliable method, as long as the board is held straight upright. If you have a compound miter saw, you can also make a bevel cut with the board laid flat, as shown (**2**). This may be your only option if you have a sliding miter saw. If you cut this way, be sure to clasp the board very tightly, because it can wander.

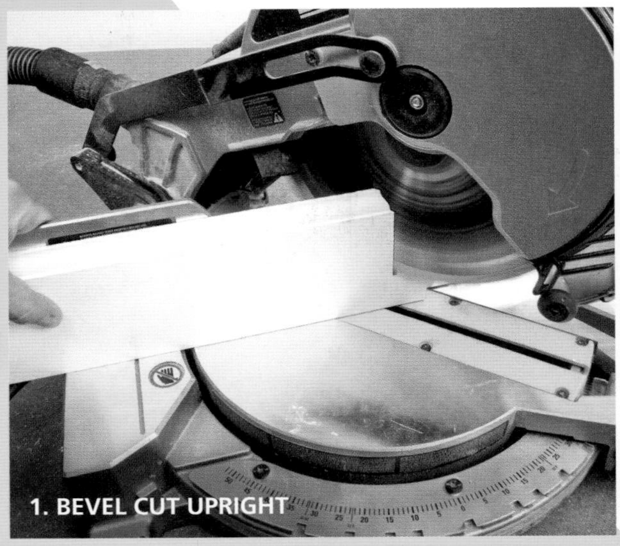

1. BEVEL CUT UPRIGHT

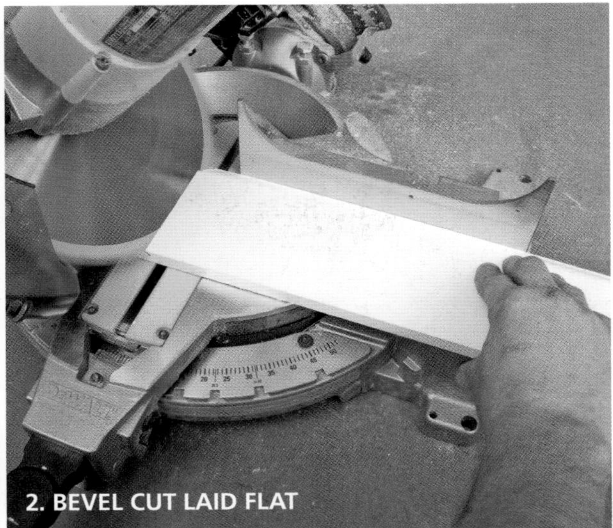

2. BEVEL CUT LAID FLAT

1. SET THE ANGLE.

2. TURN THE CAM CLAMP.

3. MAKE THE MITER CUT.

Hand Miter Saws

A good-quality hand miter saw (sometimes called a "contractor" model) can make surprisingly accurate trim cuts. Cuts will take more time to produce than if you used a power miter saw, but cutting produces only a small amount of dust that collects near the saw, so you don't need a dust collector.

Before you actually cut trim pieces for installation, make test cuts and check them for accuracy as you would for a power miter saw (p. 56). On a good-quality hand miter saw the angle can be set easily and securely (**1**). To secure the board while cutting, you could just grasp it firmly with your hand, but cam-type clamps (**2**) are quick to use and hold the board more firmly. Make the cut exerting only moderate pressure; there is no need to bear down on the saw (**3**).

Using a Tablesaw

Though perhaps not quite as easy to use for cutting trim boards, a tablesaw can accurately crosscut, miter, and bevel trim if it has a good fine-cutting blade and reliable guides. And a tablesaw has the added advantage of being able to rip-cut boards and to cut sheets for panels, making it more versatile than a chopsaw. If you are used to using a tablesaw for these purposes and have developed successful habits, then by all means stick with it.

Take a few minutes to be sure your blade and guides are precisely set up, so you can make accurate cuts with ease. A good-quality new saw will probably be accurate right out of the box, but it's best to check. Older saws may need adjustments.

Crosscutting

A good tablesaw has a miter guide that slides in the table's groove smoothly but without wobble. This produces very accurate 90-degree crosscuts. To be sure of cuts that will make tight joints, make a test cut and check that it is square. Also make sure that the bevel—squareness through the thickness of a board—is accurate.

TEST THE BEVEL. Test your sawblade's bevel to be sure it is 90 degrees to the table. Cut a 2×4 or other fairly thick piece. Flip one of the cut pieces over; the two edges should meet tightly along their thickness. If not, you may need to slightly change the bevel.

TIP Because it can sometimes be difficult to precisely line up the sawblade with an angled line on a piece of trim, you may need to sneak up on the cut, as discussed in the tip for miter saws on p. 59.

1

2

MAKE A CROSSCUT. Adjust the blade to about ⅛ in. above the thickness of the board. Stand directly behind, so you can line up the sawblade with the cutline. Start the saw, hold the board firmly to the miter guide, and push forward smoothly (1). Once the board is fully cut by the front teeth, do not keep pushing; the back teeth could tear up the cut a bit. Instead, pull the cut board away from the blade (2).

TIP Crosscutting a very long board can be difficult to do with a tablesaw, because most of the board will hang off the table. If you have an abutting worktable or nearby support at the same height as the saw's table, that will certainly help. An assistant may be able to help, as long as he or she simply supports the board from underneath and does not grab it.

CHECK THE MITER. Make 45-degree test cuts, practicing until you can make them with ease. Check the miter with a sliding square. If the miter is inaccurate, adjust the miter guide; you may need to loosen a couple of screws and a hex nut, adjust, and retighten.

CUT THE MITER. Squeeze the board against the miter guide with both hands and slide smoothly through the saw. Pull the cut board away from the blade once the front teeth complete the cut.

Back Cutting

Wherever you need to butt a trim board against a wall or a perpendicular trim piece, back cutting—also called undercutting—can produce a tighter, neater joint. The idea is to make the exposed side of the board just slightly longer than the back side, which will not be visible once the board is installed.

It's common to back cut about a 3- to 5-degree bevel. If the top edge of the back side will show, 1 or 2 degrees is a good choice.

There are various ways to back cut. When cutting with a miter saw, you could slightly adjust the saw's bevel (if it is a compound miter saw), or raise up the board as shown (**1**). Or after making the cut you can use a belt sander, plane, or Surform tool, as shown (**2**), to shave the back side.

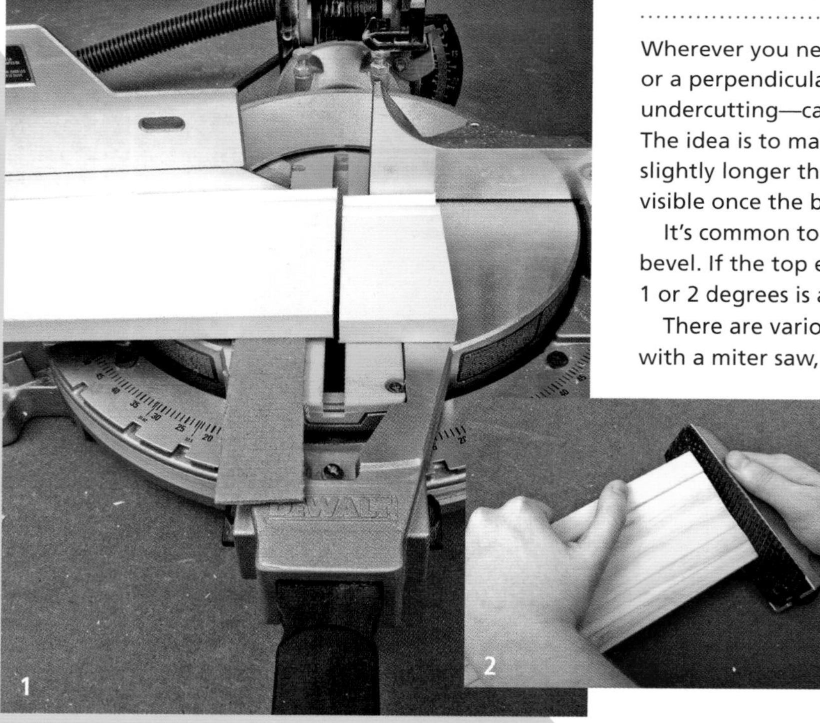

TIP **WIDEN A MITER GUIDE** A typical miter guide is only about 5 in. wide, making it difficult to keep a board from wobbling while making a crosscut. If you attach a piece of plywood about 10 in. long to the guide as shown, it will be easier to hold the trim board stable during a cut.

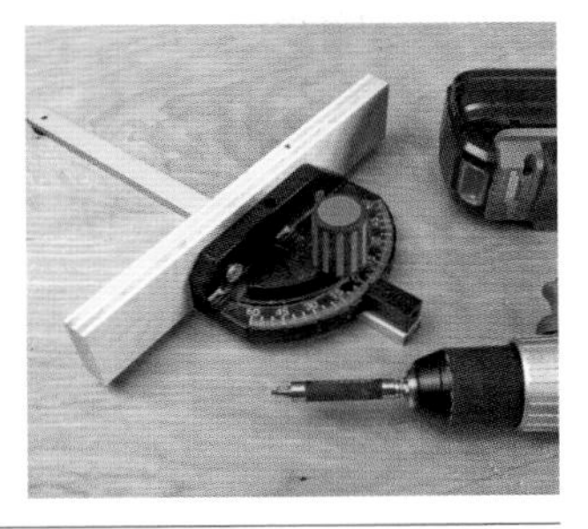

Miter cutting

Cutting a board at 45 degrees or another miter angle on a tablesaw calls for careful lining up of the blade with the cut mark. Also, hold the board very firmly against the miter guide, because it has a tendency to slide while being cut (see the top photos on the facing page).

Bevel cuts

Some carpenters find that bevel cutting is easier with a tablesaw than with a miter saw. That is, as long as the blade is adjusted to the precise bevel angle (photos below).

TIP **ADJUSTING BLADE HEIGHT FOR A CLEAN CUT** Tearout is not usually a problem when crosscutting along the grain of a board or plywood sheet, but it often occurs when cutting across the grain. The easiest way to minimize tearout is to adjust the sawblade's height to just barely above the thickness of the sheet, so part but not all of the blade tips emerge above the sheet.

CHECK THE BEVEL. Turn the crank to adjust the saw's bevel to 45 degrees or the desired angle. Make a test cut on a fairly thick piece, and check the bevel with a sliding square. If the bevel cut is inaccurate, turn the crank to change the bevel—and draw a mark on your bevel guide for future reference—or follow manufacturer's instructions to adjust the saw's bevel.

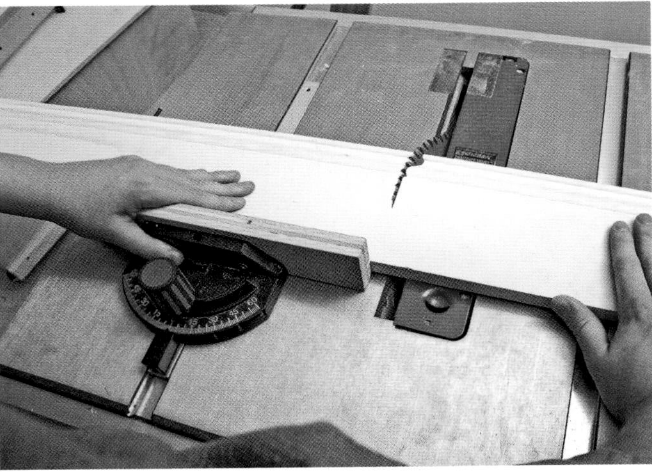

CUT THE BEVEL. Hold the board against the miter guide, carefully line up the blade with the cutline, and push through smoothly. Pull the cut piece away as soon as it is cut through.

Ripping

Rip-cutting boards to length is not a common operation for trim carpentry, but you may need to rip a board to length after shaping it with a router, or cut largish plywood sheets. A tablesaw is the ideal tool for these operations.

To make the cut, first position and tighten the rip fence the correct distance away from the blade. Then set up a featherboard. You can make cuts without a featherboard, but it is easy to install one and it has advantages: It ensures the work will not wander away from the fence, and it reduces the risk of kickback.

TIP **SET THE FENCE TO AVOID BINDING AND KICKBACK** **Carefully measure the distance between the rip fence and the blade tips, both in front and on the back side of the blade. The distance should be the same, or the back blade tips should be 1/32 in. farther from the fence. If the back blade tips are closer than the front ones, the work will bind as you cut and could kick the cut piece back like a bullet.**

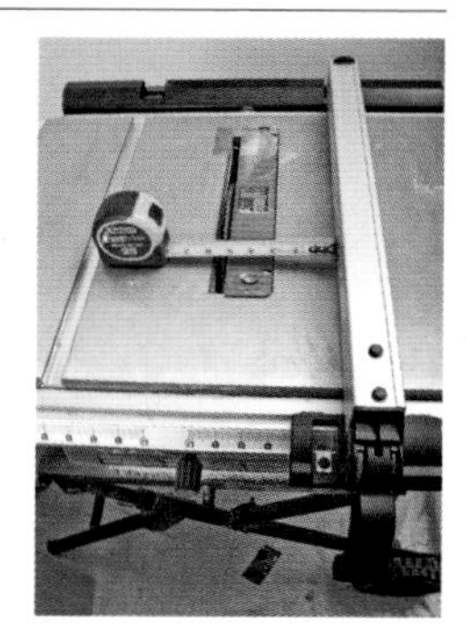

Blade guard removed for clarity.

USE A FEATHERBOARD FOR SAFETY. Position and tighten the featherboard so it is fairly snug, but not too tight.

TIP **As much as possible, push in a long, continuous motion to avoid stopping and starting when rip-cutting. Stops and starts often create irregularities in the cut.**

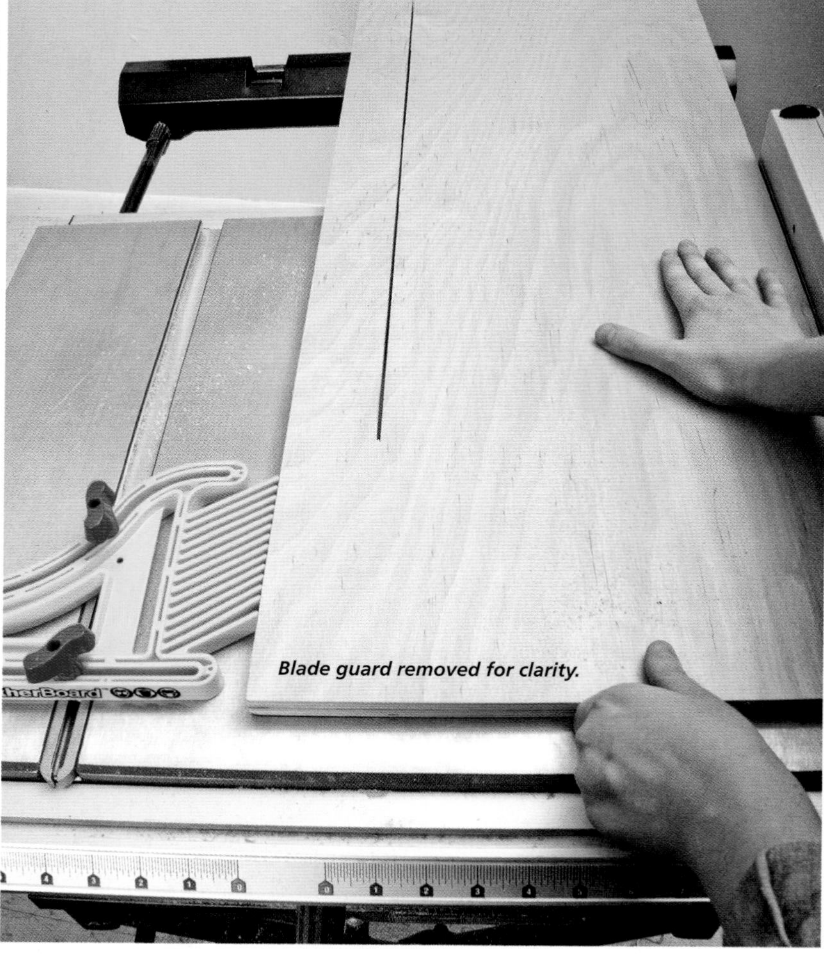

Blade guard removed for clarity.

MAKE THE RIP CUT. Turn on the saw and push the work through smoothly. Unlike with a crosscut, you should continue pushing the sheet all the way through the blade.

Cutting Trim with a Circular Saw

Using a circular saw for trimwork is not ideal, but with practice and a couple of guides it can work well. If the waste side of the cut is short, support the board so the waste falls away. If the waste side is long, support the board on both sides.

1. Use a square as a guide for square cuts. Clamp the square firmly and pass the saw's plate against it as you make the cut.

2. For rip cuts, clamp a long metal guide at both ends.

3. Make any miter cuts on the back side of the board. Clamp a square as a guide and cut the board a bit long; then straight-cut the bottom to the exact length.

1. CROSSCUTTING

2. RIP-CUTTING

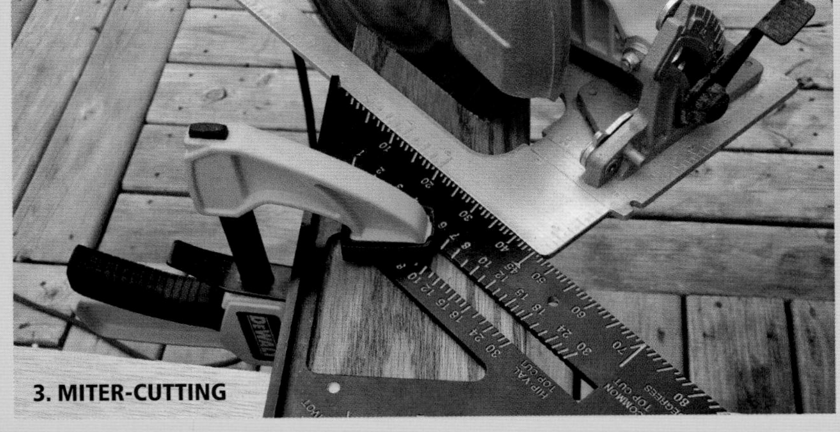

3. MITER-CUTTING

Nailing

Trim is usually attached with nails rather than screws. The goal is to attach firmly and as close to invisibly as possible. This is usually done with finish nails, which have small heads that add a bit of grabbing power while producing only small holes.

Hand nailing

If you have only a small amount of trimwork to do, hand nailing can work just fine. However, one mis-hit will create a smile- or frown-shaped dent that is difficult to mend. So practice on scrap pieces that are positioned in the same orientation as the boards you will attach—usually on a wall at various heights—until you feel confident of your skills.

Practice until you can swing a hammer with a loose wrist and a relaxed motion, so it "snaps" at the end of the blow. If you hold your wrist rigid, you'll get sore quickly, and you'll be more likely to mis-hit.

TIP **When possible, drive nails into a part of the board where the nail head will not be highly visible.**

TIP **Finish nails should penetrate the framing by 1 in. or so. If you have ½-in. drywall and a trim board that is ⅜ in. thick, use nails that are at least 2 in. long; they will penetrate the framing by 1⅛ in. If you are working on an old plaster wall that may be 1 in. thick (counting the plaster and the wood lath, which cannot be relied on to hold a nail), use longer nails.**

1 START THE NAIL. Hold a trim nail in position with your thumb and finger. Tap it with a hammer once or twice until it can stand on its own.

2 DRIVE THE NAIL. Get your body into a comfortable position—not too close to the nail—so you can swing freely. Drive the nail in two or three strokes until it is ⅛ in. or so proud of the board. Do not try to drive it all the way flush, or you will likely dent the board.

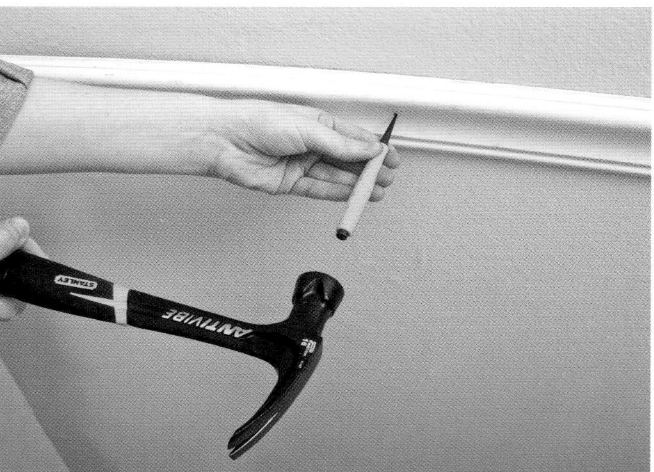

3 SET THE NAIL. Finish with a nail set sized to fit into the cup of the nail head so it won't slip out. Drive the nail head about ⅛ in. into the board. You'll fill and sand the resulting hole later.

Power nailing

If you have a power nailer, by all means use it even if you have only a few nails to drive. It really takes only a few minutes to set up, and if it saves you from making a dent in the wood, the setup time will be well spent.

TIP **NAILING AT THE FLOOR** **Different heights call for different nailing postures and swinging motions; learn how to do them comfortably. When nailing baseboard, you may be tempted to kneel down and nail carefully. Instead, develop a relaxed swing from a standing position.**

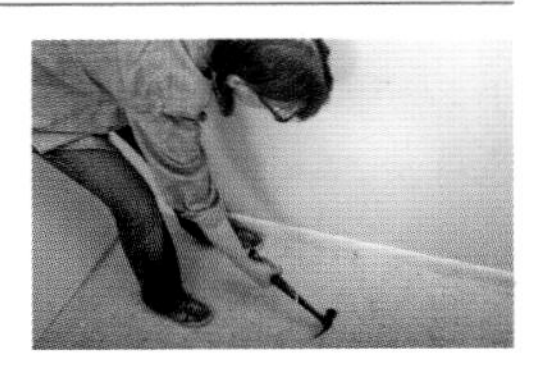

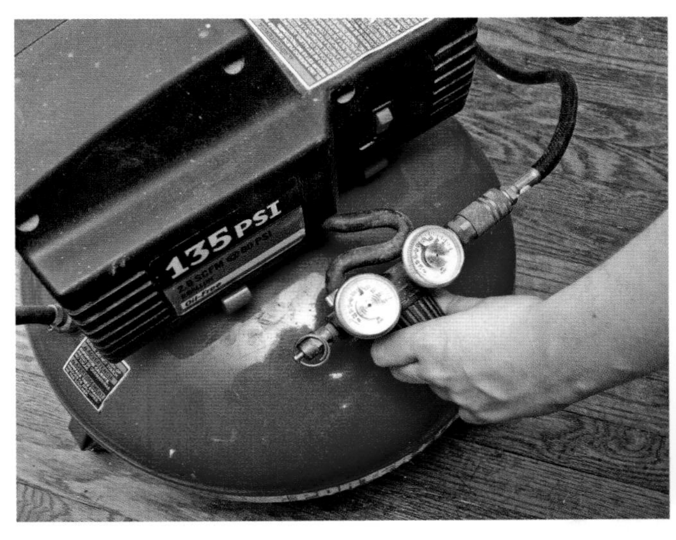

ADJUST THE PRESSURE. Turn the air compressor on and allow it to run and come up to full pressure and shut itself off. Turn regulator pressure to zero, then connect the air hose and nailer. Adjust air pressure, and drive nails into scrap pieces to find the best pressure setting that will sink the nail heads slightly below the surface of the board.

POWER NAIL. Find a spot where the gun's tip can comfortably rest without slipping. Press the tip all the way in, push with medium pressure, and squeeze the trigger to drive a nail.

Find Framing Members

Nails should be driven into wall studs or they won't hold. (In unusual circumstances where you cannot find an underlying framing member, you may glue a board in place and drive nails at angles into the drywall only to temporarily hold the board until the adhesive sets.) A stud finder like the one shown here can point the way. Position it in the center of the located stud and a plumb laser beam will point the way to the right nailing spot.

If your walls are lath and plaster, a cheap stud finder will not be able to locate the studs because it will sense all the wood lath in front. Buy a finder that can be adjusted for depth, so it can see the stud behind the lath.

Making a Coped Joint

Where two trim boards meet at an inside corner, you could cut both pieces at 45 degrees and butt them together; the result may be a fairly acceptable joint, but it will rarely be very tight, especially if the walls are imperfectly square (as they often are). A much more reliable method is to make a coped joint.

Coping a joint may appear difficult, because you have to cut along a curved profile with precision. However, with a bit of practice and a good coping saw you'll get the knack pretty quickly.

TIP **Whenever possible, cut the coped end on a board that is longer than it needs to be and then cut the board to length after you are sure of the fit. That way, if you make a coping mistake, you can try again.**

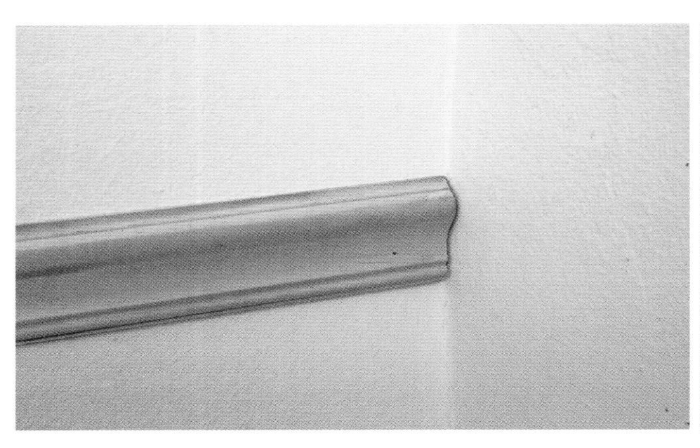

1 **BUTT-CUT THE FIRST BOARD. Cut the first piece at a simple 90 degrees and install it tight to the wall.**

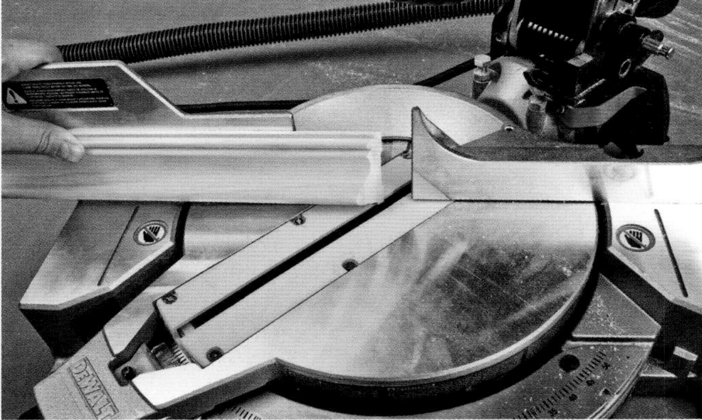

2 **REVERSE MITER THE SECOND BOARD. Cut the second piece, which will be coped, at a reverse 45-degree miter, so that the exposed cut edge faces outward.**

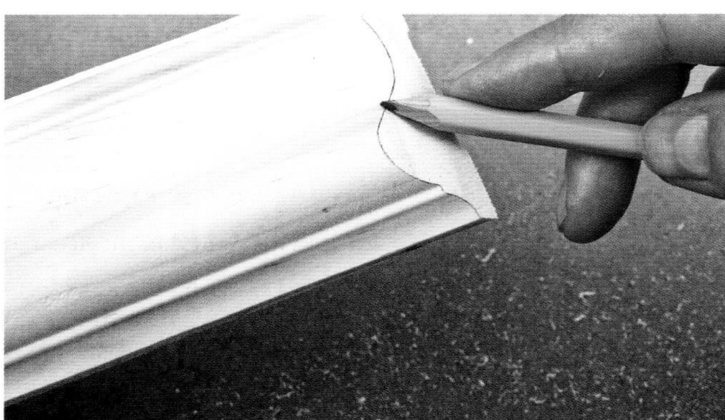

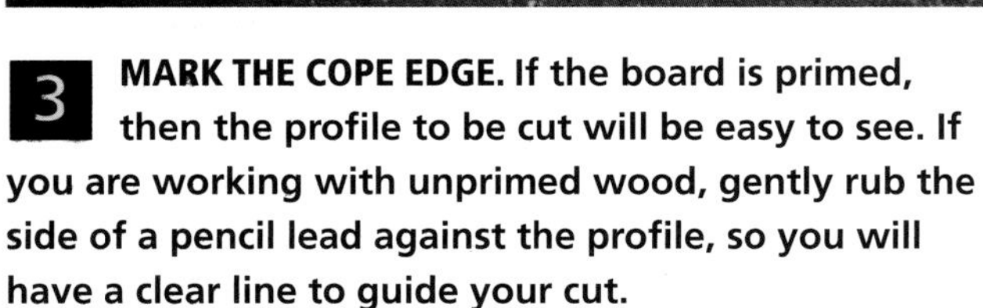

3 **MARK THE COPE EDGE. If the board is primed, then the profile to be cut will be easy to see. If you are working with unprimed wood, gently rub the side of a pencil lead against the profile, so you will have a clear line to guide your cut.**

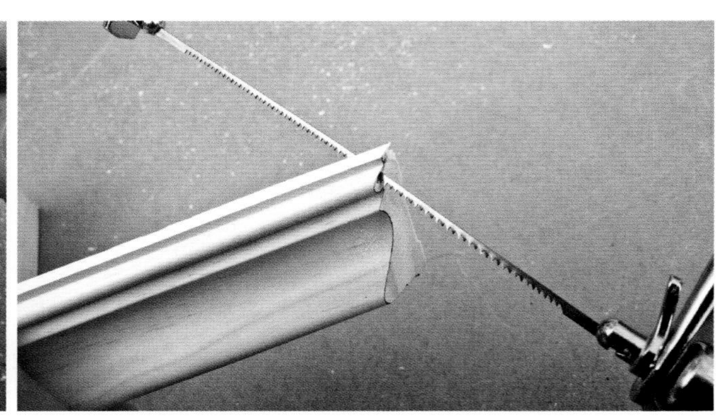

4 **START THE COPE CUT. Using a coping saw, begin the cope cut at the top or bottom. Be sure to back cut: Hold the saw so you are cutting at more than a 90-degree angle in relation to the face of the board. (You want the face of the cut to extend farther than the rest, so the profile can snug against the other board.)**

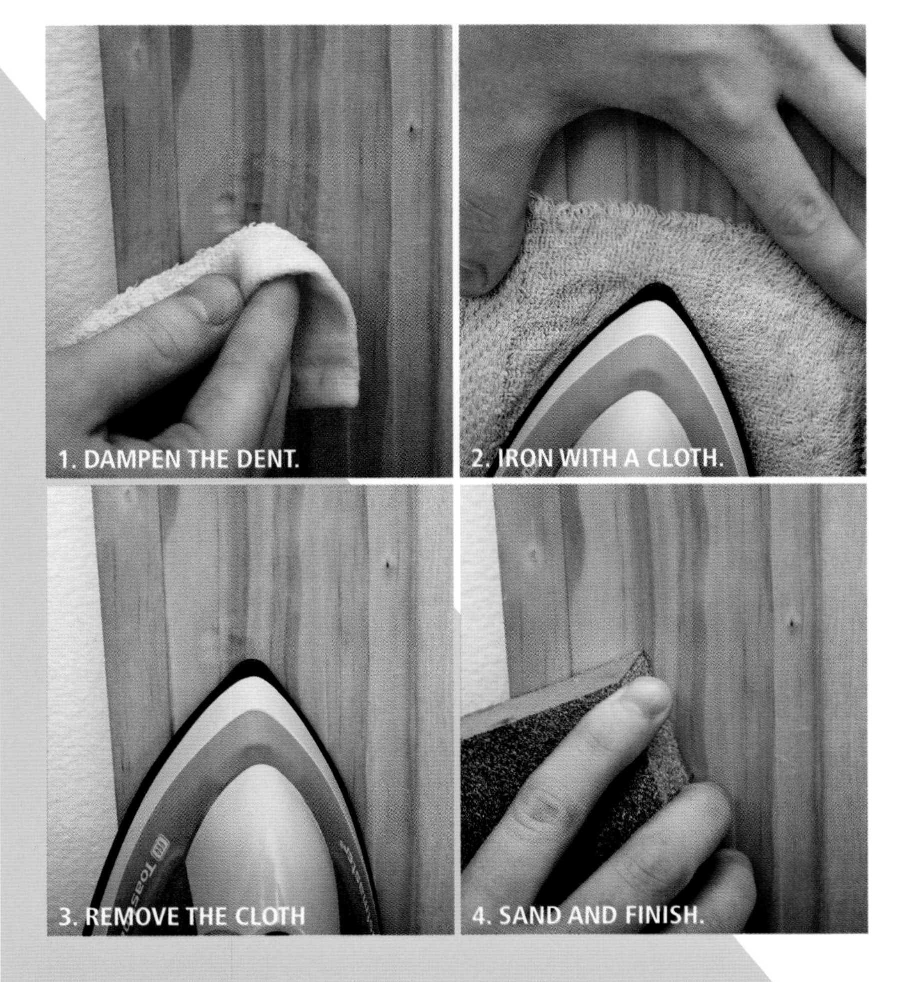

Repairing a Dent

Especially if you hand-nail and especially if the trim is softwood, dents can happen. Fixing one is surprisingly easy if the wood is bare.

1. Turn a household iron to a high setting and allow it to heat up. Wet a cloth with water and press to dampen the dent.

2. Place a cloth over the dent and press the dented area with the hot iron. This will cause the indentation to rise up to the level of the surrounding wood.

3. If the dent has not disappeared, iron carefully without the cloth.

4. The moist heat may cause the dent and the surrounding area to swell a bit. Allow to dry, sand smooth, and apply finish.

5 CUT SHY OF THE LINE. Take care to cut against the line, or slightly to the waste side of it, but don't cross over the line and cut too deep. (You can fine-tune the cut later with a knife or rasp.) The coping saw will turn surprisingly tight curves, but when you get to a curve that is too tight to follow, pull the saw out and cut from the other side.

6 **CUT IN SECTIONS.** If a profile has several curves, it's usually easiest to cut in sections. When you get to a very detailed area, like the very bottom of this piece, don't attempt to cut the tiny angles with the saw; just cut across.

7 **CLEAN UP THE POINT.** Once the section is removed, you can go back and make the tiny angled cuts.

8 **FINE-TUNE WITH A KNIFE.** The cut won't be perfect, but fine-tuning it will take only a few minutes. Use a knife to cut into small crevices and remove burrs. You may also need to cut with a knife to remove any places that have not been back-cut enough, thus preventing the coped line from contacting the other board.

9 **FINE-TUNE WITH A RASP.** You may also use a rasp to bring the cut precisely up to the cutline. Always work holding the tool so it back-cuts.

10 **INSTALL THE COPED BOARD.** Press the cope-cut piece against the straight-cut board you installed in Step 1. You may need to further slice or rasp away some places. Once the cut fits well, cut the board to length and install it.

TIP If you cut too deeply and cross the cutline, it's usually best to recut the reverse miter and start over.

Making a Return

If a trim piece ends in the middle of a wall, rather than butting against a wall corner or against another piece of trim, simply straight-cutting it will create an unfinished look. Instead, cut and install a return, so the board visually makes a turn and dead-ends into a wall. Returns are often used for window aprons (see p. 132), or for chair or plate rail that stops short of a wall corner.

TIP **It may take some trial-and-error cutting to make the perfect return piece. Fortunately, this doesn't waste much trim board length.**

1 **CUT THE LONG PIECE. Cut the molding to the desired length, with a 45-degree bevel facing the wall.**

2 **CUT A SCRAP PIECE. Also cut a 45-degree bevel on a short scrap piece, to be used for the return. Note that the return piece is cut at a mirror image of the trim piece's cut; position the board upside down to do this.**

3 **MARK THE BACK OF THE BOARD. The return piece will be cut at the length of the bevel. Mark for the cut by rubbing a pencil against the cut, omitting the thinner part of the board, as shown.**

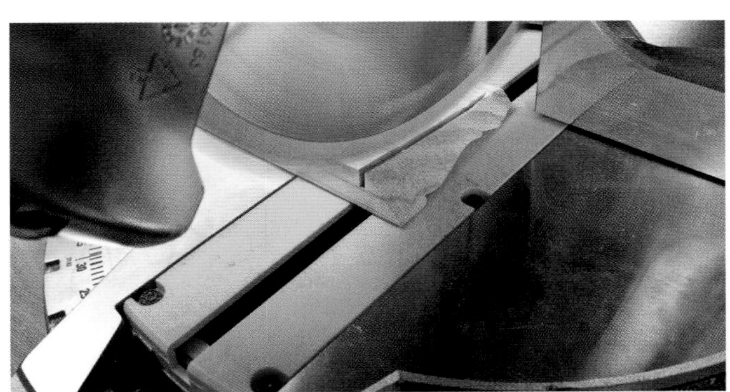

4 **CUT THE RETURN. Set the saw to 90 degrees, and cut along the lines to produce a small return piece that is just as long as the 45-degree bevel. This should produce a piece that is as long as the thickness of the trim board.**

5 **NAIL AND GLUE. Test that the return piece fits neatly, spread some glue on its back, and drive a couple of pin nails to hold it in place while the glue dries.**

Planing

Use a plane to straighten out a board's edge, or to remove thin wood shavings. When doing trim work, a plane is often used to shave a jamb or framing that protrudes slightly proud of the wall surface, to enable the trim to sit flat on the wall. You may also use a plane to cut along a curved scribed line, so that, for instance, a piece of molding will neatly follow a curved floor line.

TIP **Planing is easiest when you are going in the direction of the grain. That is, unless the grain is perfectly parallel with the board's edge, the grain lines should rise toward the edge in the direction you are planning, so the blade will not dig in.**

POWER PLANING. **A power plane makes quick work of smoothing and straightening. To bring a protruding window or door jamb flush with the wall, hold the plane with the non-cutting edge of the base held against the wall as you cut.**

BLOCK PLANE. **Adjust a block plane so its blade just barely protrudes below the sole, and tighten it. Some planes have adjusting knobs in the front and back. An inexpensive model like this one adjusts by loosening the blade, sliding it forward or back, and tightening. Make test cuts with the plane and readjust until it slices easily, with little downward pressure.**

PLANING TRIM. To plane a board's edge, clamp the board or position it so it will not wobble or slide as you work. Press down with only moderate pressure and push the plane with long, smooth strokes. If the going gets rough and the blade sticks, try planing in the other direction.

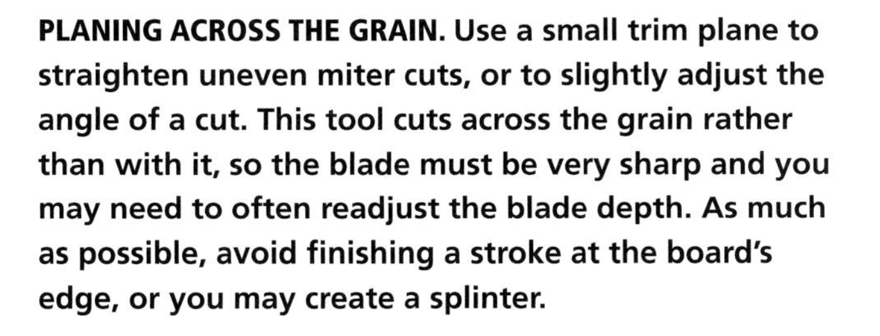

PLANING ACROSS THE GRAIN. Use a small trim plane to straighten uneven miter cuts, or to slightly adjust the angle of a cut. This tool cuts across the grain rather than with it, so the blade must be very sharp and you may need to often readjust the blade depth. As much as possible, avoid finishing a stroke at the board's edge, or you may create a splinter.

ROUGH SMOOTHING. A Surform® tool is easy to use and requires no adjusting, but the results are less than smooth. Use a Surform when you need to remove material on the back of a board, where it will not show.

Biscuit Joinery

With a biscuit joiner (also called a plate joiner) you can easily and firmly attach boards or sheets on the same plane, with no visible fastener heads. The joiner cuts a slot into each board, into which fits a wafer-shaped biscuit. The holes and wafers receive plenty of glue, making for a strong joint with no visible fasteners.

Biscuits are sometimes used even when pieces will be nailed or screwed together, to ensure that faces will be precisely aligned, avoiding the problem of one piece sticking out past the other.

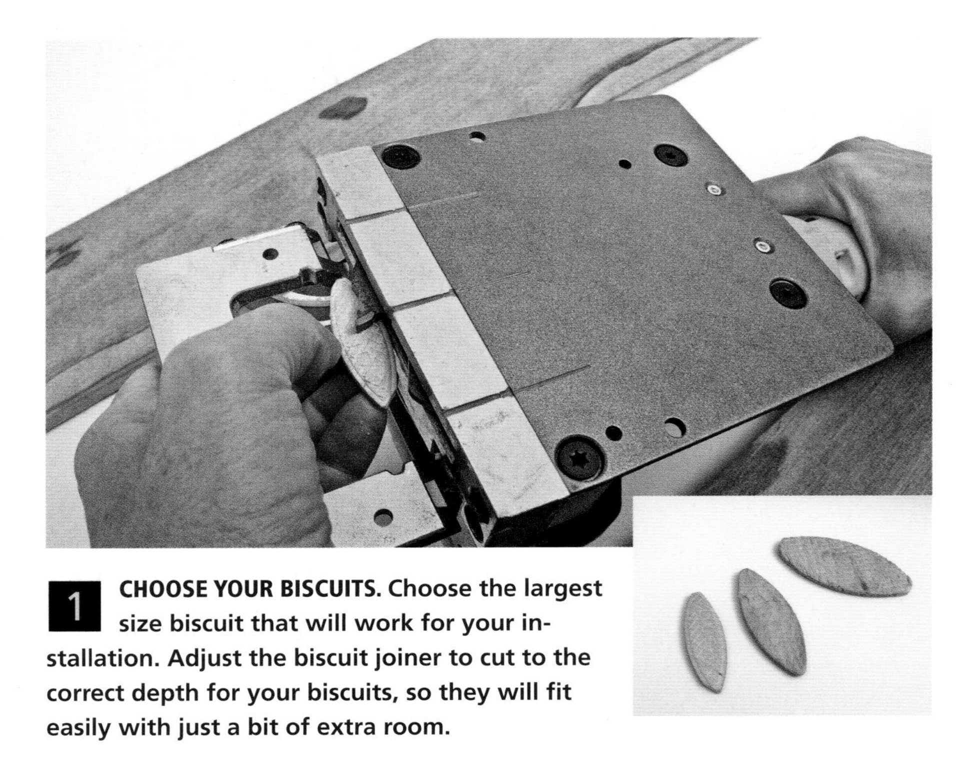

1 **CHOOSE YOUR BISCUITS.** Choose the largest size biscuit that will work for your installation. Adjust the biscuit joiner to cut to the correct depth for your biscuits, so they will fit easily with just a bit of extra room.

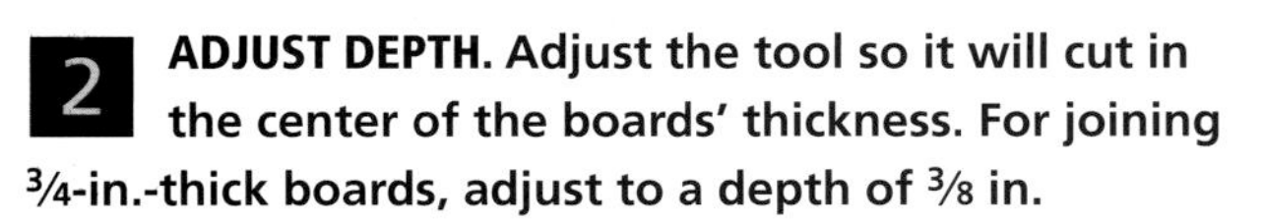

2 **ADJUST DEPTH.** Adjust the tool so it will cut in the center of the boards' thickness. For joining ¾-in.-thick boards, adjust to a depth of ⅜ in.

3 **MARK SLOT POSITIONS.** Hold the two pieces together, aligned as you want them to be fastened. At the centers of where you want to place biscuits, draw lines that span from board to board.

TIP Wood glue sets up pretty quickly, so be sure you have all your ducks in a row before applying glue. Ideally, you should clamp the boards together within 2 minutes of beginning to glue.

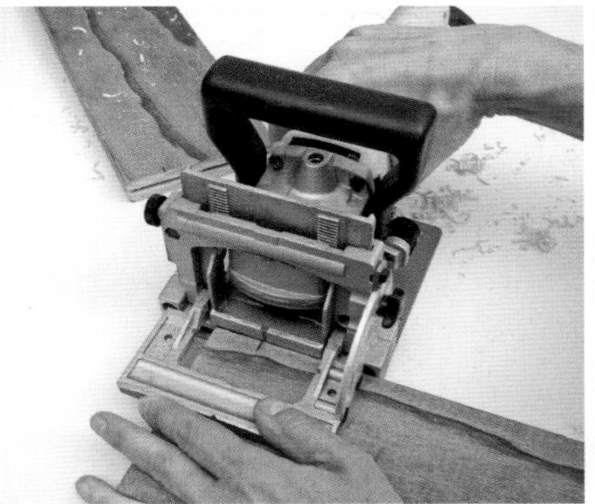

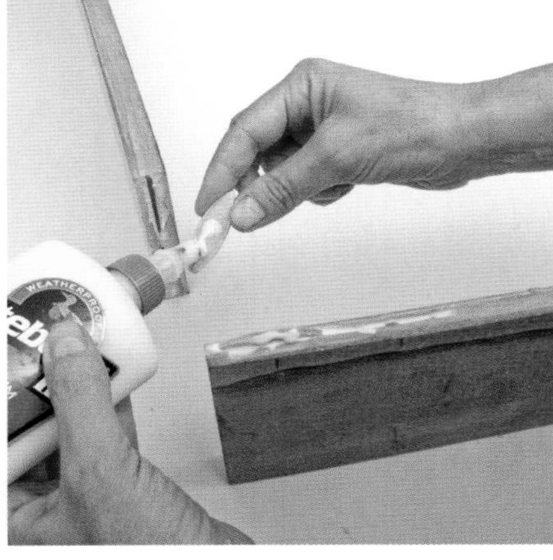

4 **CUT THE SLOTS.** Place the biscuit joiner's front plate solidly on top of one of the pieces, align its center guide with a cutline, turn on the motor, and press firmly to cut a slot. Repeat for the other piece.

5 **ASSEMBLE AND GLUE.** Assemble the two pieces with biscuits so you are sure they will fit. Disassemble, and apply glue to the slots and to the biscuits.

6 **ALIGN AND CLAMP.** Clamp the pieces together with moderate pressure, check for alignment and tap one board to the side as needed to achieve perfection, and tighten the clamps firmly. Allow the glue to dry.

Biscuits for Miter Joints

You can reinforce a miter joint with biscuits, but because biscuits are fairly wide, this works only for large casing pieces. Cut the miters (or, in the case of Craftsman casing, cut the pieces at 90 degrees), and check the fit. Working on a flat table, cut the biscuit slots. Assemble the pieces, with the biscuits, on the wall, and drive nails or screws to hold them together. The biscuits add extra strength and ensure that the pieces are on the same plane.

Using a Router to Make Custom Trim

You may find yourself trying to match existing trim in an older building. Often trim pieces with these profiles cannot be found in home centers, lumberyards, or even from online sources.

Fortunately, a wide assortment of router bits is available that allows you to cut decorative profiles onto standard dimensional lumber boards. The result will likely not be a perfect match for the old trim, but it can be close enough that the difference will not be noticeable—especially if the old and new boards are not placed side by side.

Freehand routing

Using a router bit with a rolling guide bearing, you can cut profiles freehand, without a table. Once adjusted correctly, the guide bearing will keep the bit from digging in too deeply, so you can maintain an even line—as long as the board itself is straight and free of edge indentations.

ROUTER BITS. An assortment of router bits can be used, perhaps in combination, to make a great many profiles.

1 INSERT THE BIT. The shaft will be ¼ in. or ½ in. diameter; some routers accept only one size, whereas others can use either size. Push the bit all the way in, then pull it back about ⅛ in. Tighten the bit firmly, using two wrenches (as shown) or by pushing in a collet and using one wrench.

2 SET THE ROUTER DEPTH. Depending on the type of bit, the depth can change not only the size of the profile but also its shape. Experiment with various depths, cutting scrap pieces, until you achieve the profile you desire.

3 ROUT THE PROFILE. Place the board on a flat surface and test that it will not slide while you work. When making a small piece, like the one shown here, clamp it in place. Exert even pressure against the board and push against the turn of the blade—that is, move in the direction that is most difficult to push. As long as you keep the router's base firmly flat on the board, the bit will not overcut. You can rerout areas that are incompletely cut.

TIP If you need to precisely match existing trim that is not available for purchase, some mills and some full-service lumberyards can produce the pieces for you, working off a sample that you give them. Generally, the cost of setting up the "knives" to make these pieces will be about $100; you'll need to pay for the pieces by the foot on top of that. If you have a lot of trim to match, this can be money well spent.

Using a Router Table

If you have a router table, you can use bits that either have or do not have a guide bearing, as the table's fence will keep the bit from cutting too deeply. A router table may be a simple shop-made affair—a table with a hole in it for the router and a fence that is simply a straightedge clamped to it. Or, like the table shown here, it may have easily adjustable guides and clamps.

1. Because the router is attached to the underside of the table, you will adjust the bit's height in order to adjust the depth of the cut.

2. The position of the fence will also help determine the shape of the profile.

3. Press the board against the fence and push it through from left to right. You can press the board against the fence with your hands, or use a featherboard to ensure that it is tight to the fence.

1. ADJUST THE HEIGHT.

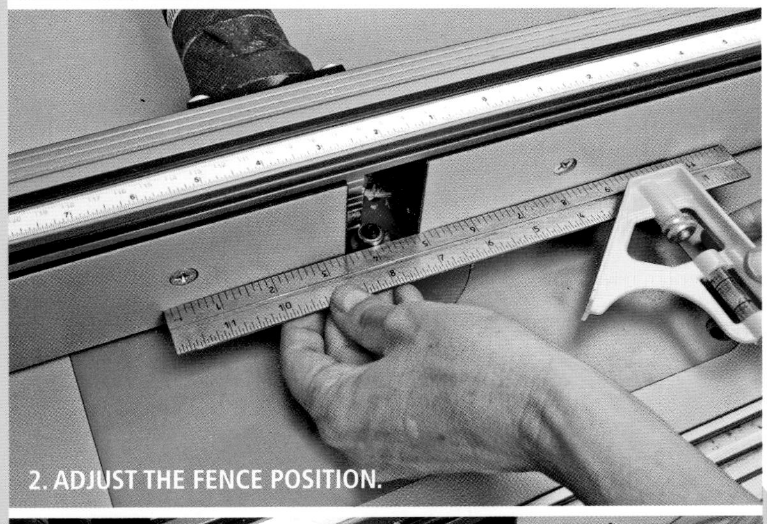

2. ADJUST THE FENCE POSITION.

3. ROUT THE BOARD.

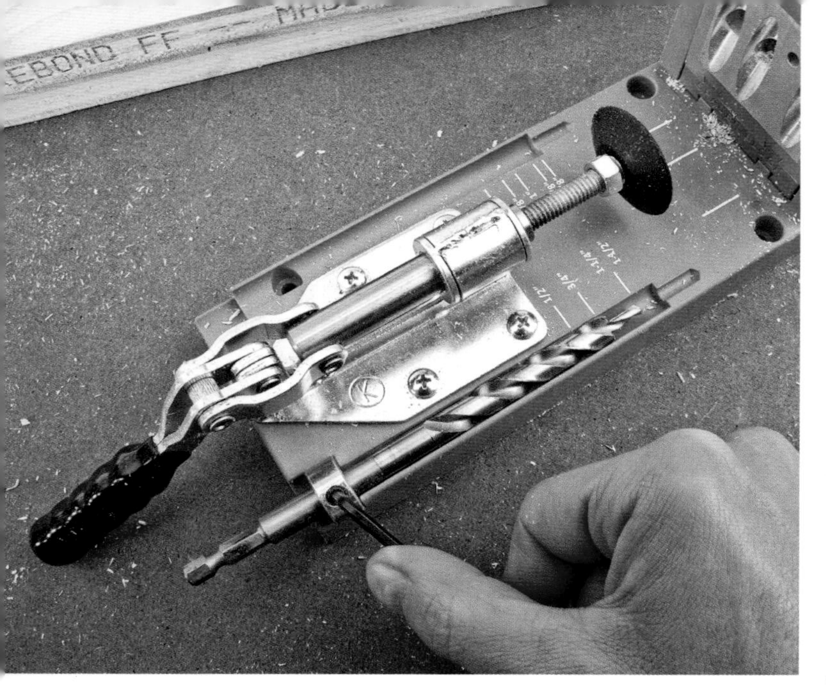

1 SET THE DRILL DEPTH. Using the depth guide, tighten the drill bit's collet so it will drill to the correct depth—in this case, for material that is ¾ in. thick.

Joining with Pocket Screws

Pocket screws allow you to quickly and securely join boards and plywood pieces, so they are useful when making wall panels, jambs, and other installations that require a frame. The system creates large holes, which are usually placed in the back, where they will not be seen (though they can be filled with special plugs made for the purpose).

Pocket systems work only when you drill holes that are precisely the right thickness and depth and fasten with pocket screws that match the depth thread coarseness. Use coarse-thread screws to attach to softwood and plywood, and fine-thread screws when attaching to hardwood. When fastening pieces that are ¾ in. thick, use 1¼-in.-long pocket screws.

The steps here show fastening 1× boards at their ends, on the same plane. Other clamps allow you to fasten boards at right angles.

TIP **Pocket screws make strong joints by themselves, but some people apply a bit of glue to the joint before driving the screws, for extra hold.**

2 DRILL HOLES. Clamp the board in a special jig and drill holes for the screws. There is no need to measure exactly for the screw positions. The drill will penetrate near to the end of the board, but not all the way through.

3 DRIVE SCREWS. Clamp the two pieces firmly together so they cannot move while you drive screws. Use the long screwdriver bit to drive the screws. If you don't get a strong joint, you may need to readjust the depth of the drill bit.

1. APPLY A LIGHT COAT OF STAIN.

2. QUICK DRY.

3. APPLY PAINT WITH A GOOD-QUALITY BRUSH.

Finishing

Whenever possible, apply finish or paint (or at least primer) to trim boards before cutting and installing them. Finishes are much easier to apply this way. (Once installed you may need to touch up with a paint brush.)

1. Use a rag to apply a light coat of stain. If you need a darker application, apply a generous coat of stain with a brush, wait a few minutes, then wipe off.

2. Stained boards will dry quickly if left outside in the sun or in a room with a fan blowing on them. However, take care to keep the boards away from blowing leaves, dust, or sawdust for as long as it takes them to dry.

3. If you're painting the boards, apply a generous coat of paint, spread it evenly, and then finish with long, lightly applied strokes for an even finish. You may need to apply two coats.

CHAPTER FIVE

WALL TRIM

THIS CHAPTER COVERS several types of horizontal trim attached to walls. Nearly all walls have baseboard trim at the bottom, to cover the joint between wall and floor. Baseboard (and base shoe) add detail that makes a room more interesting at the highly noticeable junction of floor and wall. Moving upward on the wall, chair rails—originally made to keep chair backs from denting walls—are often installed for their traditional charm. Farther up the wall, a plate rail, which is a narrow shelf, offers a practical display surface. And near the top of a wall, picture rail, originally designed for hanging artwork by wires to spare plaster walls from damage, is making a comeback. (Crown molding also touches the wall, but it has unique features and will be covered in a separate chapter.)

All these horizontal moldings are simple to lay out and install along their lengths, but making tight-looking corners on both the inside and outside calls for careful work, good tools, and a few skills that can be learned. This chapter will show how to get professional-looking results.

CLASSIC BASEBOARD. **Three-part base, with a shoe, pilaster, and cap.**

TIP **If you will stain rather than paint baseboard, be sure to use oak or another hardwood. Baseboard often gets bumped. Dents can be easily fixed and painted over, but dents in stained wood will be difficult to hide.**

MDF BASEBOARD. **These painted MDF base moldings are shaped to recreate the look of a pilaster plus cap.**

BUILT-UP BASEBOARD. **This example of a built-up base is assembled using base shoe, 1×6, and picture rail.**

Baseboard

Baseboard not only covers the gap between wall and floor but also forms a frame around a room that subtly dignifies the overall appearance of the room. Attaching it to the wall is simple, but getting tight joints at the inside and outside corners calls for careful planning and cutting.

Baseboard options

Base molding can be installed by itself, but it is more common to also install a base shoe at the bottom to cover any gap in the flooring; in combination, they create a neat and classic frame around the room.

In a modern home you may use simple ranch or Colonial baseboard—and then base shoe is usually added at the bottom. For a more distinctive effect, use a three-part molding, which features a wide "pilaster" molding with a shoe at the bottom and a cap at the top. If the molding will be painted, you can install single-piece MDF moldings that mimic the look of a pilaster plus a cap, and then add the shoe at the bottom.

Another option: Create your own unique base molding arrangement by stacking various types of moldings, perhaps using a 1×4 or 1×6 board as the pilaster.

TIP **If the floor is wavy, base shoe easily bends to follow its contours. The wider the base molding, the less noticeable the shoe's curves.**

Trim and carpeting

Where the floor will be carpeted, older installations often positioned the base or base shoe ½ in. or so above the floor so the carpeting could be slipped under. You may choose to do that, but nowadays it is common to install the base and shoe right on top of the subfloor; the carpet installers then use tack strips to attach the carpet tight against the molding.

If carpeting is already installed, and if it is fairly tight to the wall, you may choose to install only a base molding and skip the shoe. (Carpeting often hides the shoe anyway.) If you do need to install shoe, do not press down on the carpeting while nailing, or you will have to remove the shoe when you later replace the carpet.

TIP **As a general rule, base trim should coordinate with other nearby trim: It should not be noticeably wider or narrower than casings, and should have a similar level of detail.**

RECTANGULAR ROOM

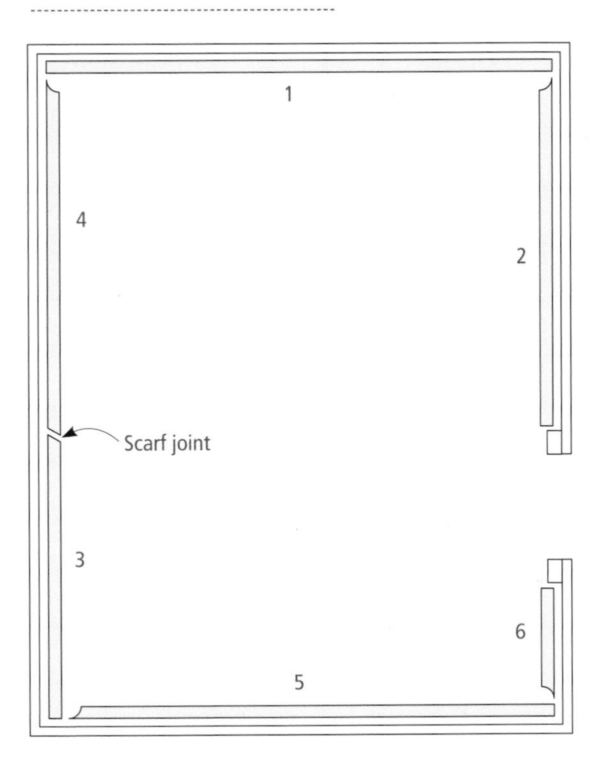

L-SHAPED ROOM

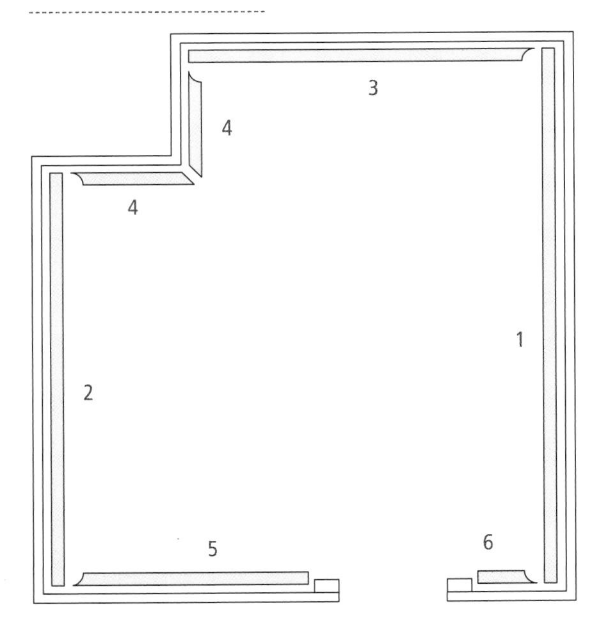

Plan the Order of Installation

When installing baseboard, there's no need to proceed around a room in one direction. Instead, plan the order of installation to minimize difficult measuring and cutting. Here are a few general rules:

- Unless absolutely necessary, avoid having to install a piece that is cope-cut on both ends; it is difficult to measure correctly for this.
- Wherever possible, plan so you can hold a piece in place and mark it for cutting, rather than using a tape measure.
- Whenever possible, plan so you will cope-cut one end of a board so it is longer than needed; then you can measure or hold it in place and mark for cutting the other end.

The drawings at left show two typical room plans; the numbers denote a recommended order of installation. Note that the majority of the pieces that run from inside corner to inside corner are butt-cut on each end.

In the bottom drawing, the two boards at the outside corner are both labeled "4" because they are best cut and installed in conjunction: Cut the coped ends, hold them in place and mark for outside miter cuts, make test miters on scrap pieces to get the angle right, and then cut the miters and hold them both in place to make any needed adjustments to the cuts (see p. 89).

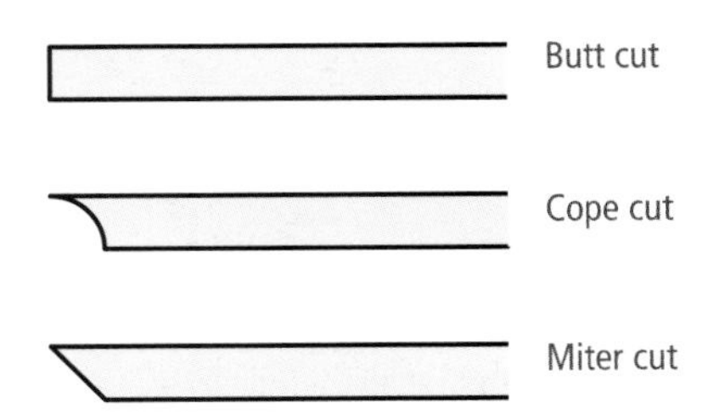

Check the walls and floor

Base moldings often cover minor waves and other irregularities in floors and walls. However, if imperfections are severe, you may need to modify the molding or one of the surfaces.

If a wall has been damaged or patched, press a piece of molding against it to make sure it can be secured flat against the wall, especially at the top where gaps will be visible. If a floor leans noticeably, check with a level; most floors are slightly out of level, which is not usually a problem, but a rise or fall of more than ½ in. over 8 ft. can be noticeable. In that case, you may need to rip-cut the base molding to bring it closer to level. Often the best solution is to go halfway toward level, because making it perfectly level could result in a molding that is clearly wider at some points than at others.

At inside corners, use a framing square to check that the walls are square to the floor, or close to it. If they are not, you may be able to fix the problem by cutting the butt-cut piece at an angle. However, if the cope-cut end of a molding piece will be at an angle, the fix is more difficult. You may be able to shim out the bottom of the butt-cut piece, but shimming the top out from the wall can result in an unsightly gap. The other alternative is to scrape the wall to make it straighter.

CHECK FOR LEVEL. If the floor is badly out of level, you may need to rip-cut the baseboard.

CHECK FOR SQUARE. Cutting the butt-cut piece of baseboard at an angle can fix this.

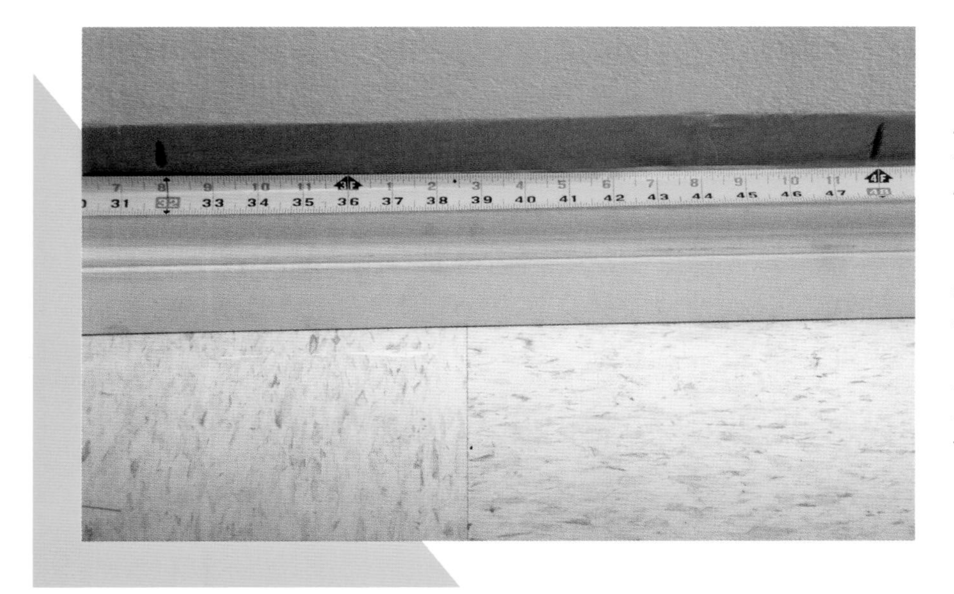

Mark for Studs

You can attach the bottom of the base molding to the wall's bottom plate, but the upper part must be attached to wall studs. Use a stud finder to locate the studs. To keep from marring a painted wall, press a strip of painter's tape just above the molding and mark the tape with stud locations.

SCRIBE AND CUT TO FIT. If you're not installing base shoe, the bottom of the baseboard must be cut to follow the contour of the floor. To scribe a line that follows the floor, mark with a carpenter's pencil held sideways, as shown. If the gap is larger than can be marked in this way, place a scrap piece of wood on the floor, hold the pencil on top of it, and slide the pencil and board together to mark a line that follows the floor. Cut the line carefully with a circular saw. (It's difficult to maintain a smooth line with a jigsaw.)

1. LAY OUT THE PREACHER.

2. MARK THE BASEBOARD.

Marking with a Preacher

The joint where baseboard meets door casing is highly visible, and the casing may not be perfectly square to the floor. To ensure a tight fit, mark with a simple guide called a *preacher*.

1. To make the preacher, place short scraps of the casing and the base molding on a small board and mark for a notch. Cut with a jigsaw.

2. To use the preacher, cut the base molding so it runs past the casing and hold the preacher as shown, so it follows the line of the casing. Mark with a pencil held so you will cut the board about ⅛ in. long, so you can bend it into place.

Installing butt-cut pieces

In most cases, you will start with pieces that are butt-cut on both ends. These should be installed snug to the corners. Where a cope-cut piece will meet at the corner, the bottom of the butt-cut piece does not need to be tight to the wall corner because it will be covered by the cope-cut piece; however, the top of the butt-cut piece will be visible, and so should be snug.

If one end of the piece will butt against a door casing, use a preacher, as shown in the sidebar on p. 85. If the piece will run from inside corner to inside corner, measure the distance and cut a piece that is a bit longer—1⁄16 in. long for a piece that is 6 ft. or shorter and 1⁄8 in. long for a longer piece. Usually it is most accurate to lay the piece flat on the miter saw's base, as shown at right, but in some cases you may choose to press it upright against the fence.

TIP **PRESS DOWN FOR A TIGHT FIT** If the base molding is not resting on the floor, place a scrap board on top and kneel on it to push the baseboard down, as shown.

1 **CUT AT 90 DEGREES.** Lower the blade and move the board until the blade is at the exact cut mark; then raise the blade, turn on the saw, and make the cut.

2 **BEND AND FIT.** Press one end of the baseboard against a corner and bend the piece slightly until it fits at the other end. Then press against the wall for a tight fit on each side.

3 **NAIL IN PLACE.** Drive nails near the bottom into the framing base and at intervals near the top into wall studs. Where possible, drive nails where their heads will not be visible to a person standing in the room.

Coping

Avoid the temptation to miter-cut pieces at inside corners; this rarely produces a neat joint. Cope-cutting is easier than it looks, and produces professional results. For more coping tips, see pp. 68–70.

> **TIP** Avoid driving trim nails with heads less than 1 in. from the end of a board, especially if the board has been cope-cut. If you must drive a nail close to the end, use a pin nailer and headless pin nails.

1 **START WITH A MITER.** Begin with a piece that is at least 2 in. longer than will be needed. Start making the cope cut by miter-cutting at 45 degrees so the part you will cope-cut is exposed.

2 **MAKE THE COPE CUT.** Using a coping saw, cut along the line created by the miter cut you just made. The coping saw's thin blade easily follows curved lines. Cut with the blade held so you slightly undercut the piece, to produce a tight fit. It's often easiest to start at the top, cut halfway down, and then cut again from the bottom up.

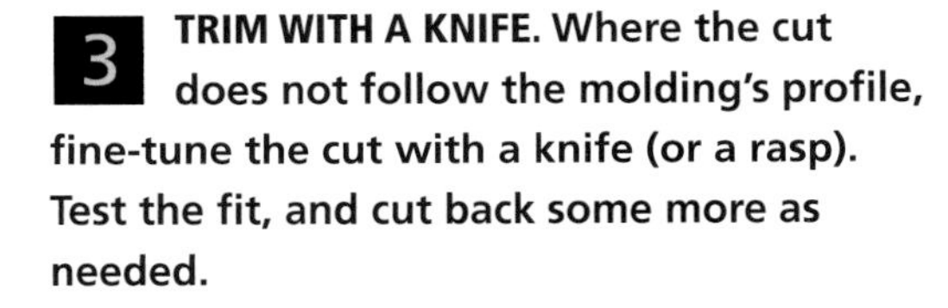

3 **TRIM WITH A KNIFE.** Where the cut does not follow the molding's profile, fine-tune the cut with a knife (or a rasp). Test the fit, and cut back some more as needed.

4 **NAIL THE COPED END.** Press the coped end in place against the adjoining piece and mark the other end for cutting to length. (This may be a simple 90-degree butt cut, a cut marked using a preacher, or a miter cut for an outside corner.) Cut the piece slightly long, so it bends tightly into place. Drive nails to attach.

Outside corners

If an outside wall corner is perfectly square and your miter saw cuts precise angles, then making a tight outside corner will be easy. Unfortunately, walls are often less than perfect. Don't settle for a corner joint that has to be filled with caulk; take the time to adjust angles for professional results.

TIP **If you are at all unsure of your miter saw's ability to produce perfect 45-degree bevels, now is the time to make some test cuts and perhaps some micro adjustments (see p. 56).**

1 CHECK FOR SQUARE. Use a framing square to check that the outside corner is at 90 degrees. If not, work with scrap pieces and make test cuts, as shown on the facing page, until you can produce a great-looking joint in the real thing.

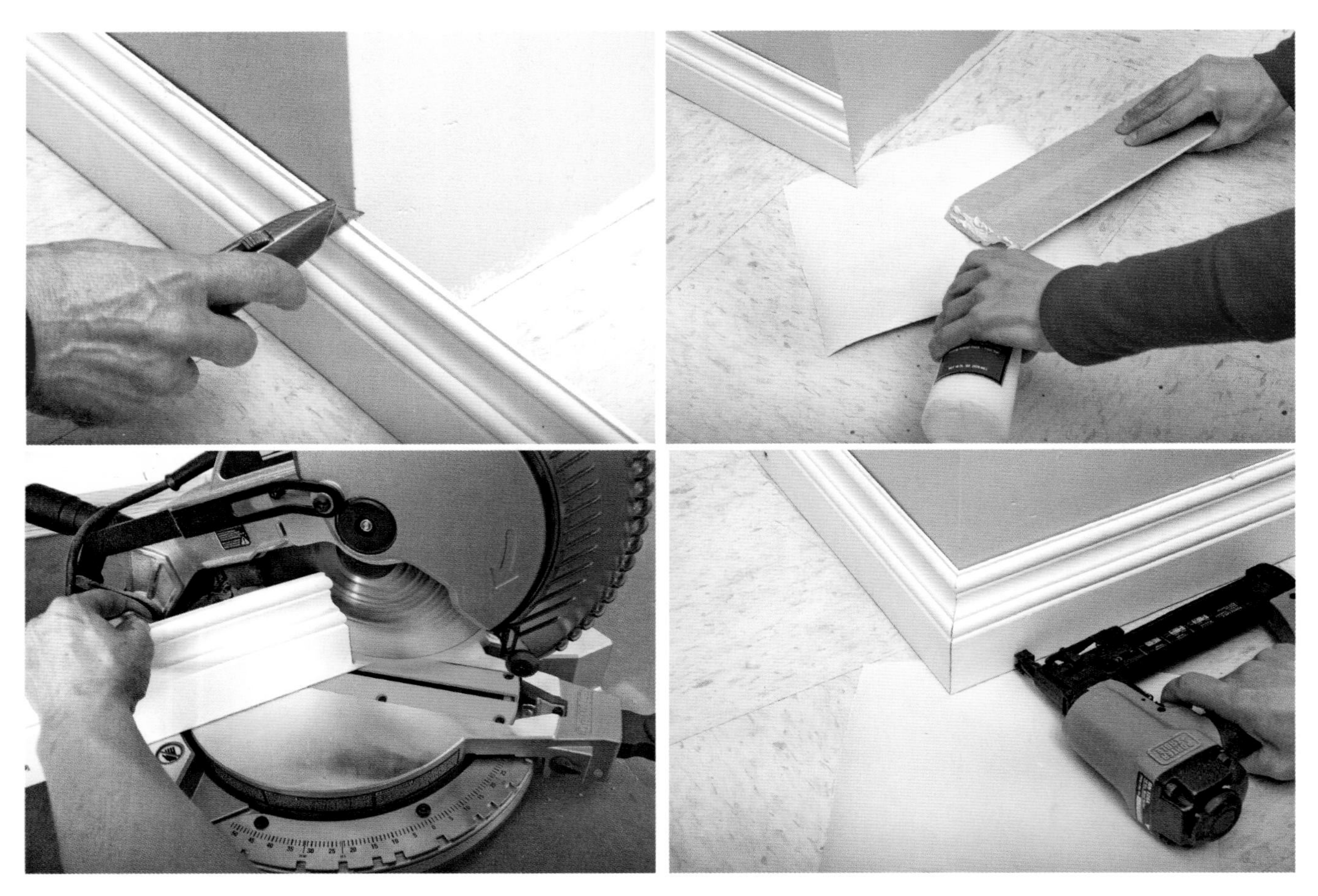

2 MARK AND CUT. Hold a baseboard in place and use a knife to mark it at the top for a cut. Cut the piece just slightly longer—about 1⁄32 in.—than the mark. Miter-cut both pieces before installing either piece.

3 GLUE AND NAIL. Press the two miter-cut pieces in place and check for fit. (If there is a problem with the fit, see the facing page for tips on making a tight joint.) Apply wood glue to one or both of the bevel-cut ends and nail the pieces in place.

If an Outside Corner Is Not Tight

If an outside corner is not tight, the problem may be addressed in several ways.

1. If the joint opens up slightly at the front (**1A**), try using a plane or a Surform tool to remove material at the back (**1B**). Cut only where it will not show.

2. Use an angle gauge to measure the angle of the corner. If it is not 90 degrees, change the angle setting on the miter saw from 45 degrees to compensate (**2A**). Or position a shim behind the piece, against the fence, to slightly change the angle (**2B**).

3. If an outside miter is only slightly open, you may choose to use a pin nailer to pull things tight. Use headless pin nails and attach them at a slight angle, as far back from the corner as possible. Drive at least one pin into each molding piece, to lock the joint in place (**3**).

1A.

1B. PLANE THE BACK.

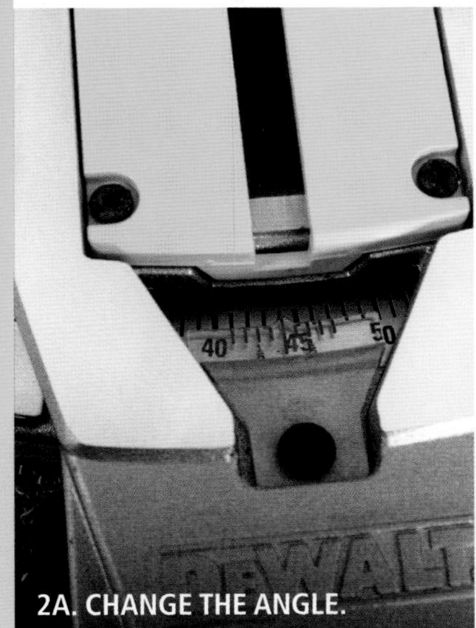

2A. CHANGE THE ANGLE.

2B.

3. PIN IT.

Scarf Joint

If a wall is longer than your longest molding piece, you'll need to splice two pieces. This is typically done with a "scarf" joint, which is less visible than simply butting two straight-cut pieces. Cut one piece at 60 degrees, with the cut side exposed, so it is 1 in. to 2 in. short of a wall stud. Attach the first piece. Cut another piece at 60 degrees, with the cut edge hidden. Test that the joint will be tight; sand or plane the back of the second piece if needed. Apply glue, and attach to the wall. Wipe away squeezed-out glue.

1. GLUE.

2. NAIL.

3. WIPE.

Installing base shoe

Base shoe widens base molding to cover a wider gap at the floor, and it completes the look of plain base molding by adding another line. Cut and install base shoe in the same sequence as for base molding (see p. 83).

NAIL BASE SHOE. Install butt-cut shoe pieces that are slightly long, so you need to bend them into place. Press the shoe down onto the floor, but do not nail to the floor. Instead, direct nails in a mostly horizontal direction, so they attach to the base molding and/or the wall framing.

> **TIP** **BASE SHOE ORIENTATION**
> **Although quarter-round molding is the same width in either direction, base shoe typically has one shorter side and one longer side. Place the longer side on the floor if you need to cover a wide gap. If covering a gap is not an issue, you may choose to position the shoe in either direction, depending on the look you prefer.**

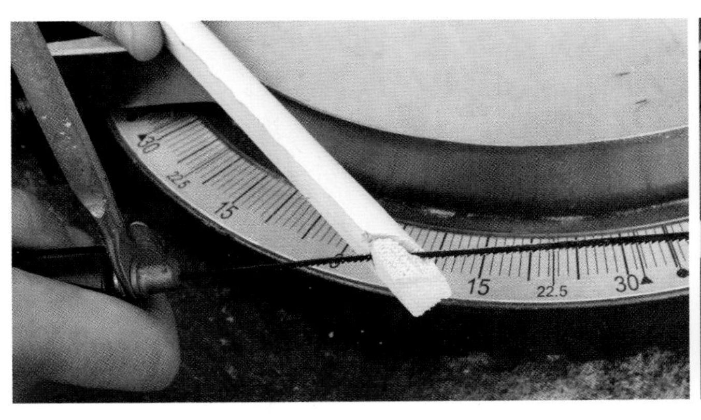

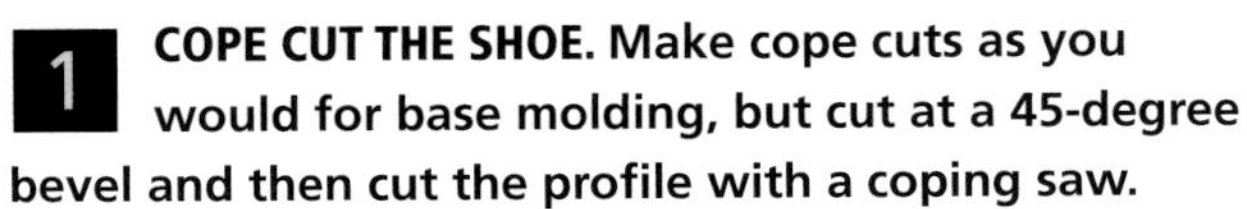

1 **COPE CUT THE SHOE.** Make cope cuts as you would for base molding, but cut at a 45-degree bevel and then cut the profile with a coping saw.

2 **TRIM THE BACK.** You'll have to trim the back of the cut pretty severely, until it forms a long point.

3 **DON'T NAIL TOO CLOSE TO THE ENDS.** Because shoe is narrow, avoid driving nails within 3 in. of the ends, to prevent splitting.

Three-part base molding

There are various types of three-part moldings. These generally add a few quirks to the installation process. Refer to the illustrations on p. 83 for a typical installation sequence in a room.

The central piece, referred to as pilaster molding, is often tall with only one profile detail. At an inside corner, first install the butt-cut piece. You could then make a cope cut on the second piece, or follow the sequence shown in the photos.

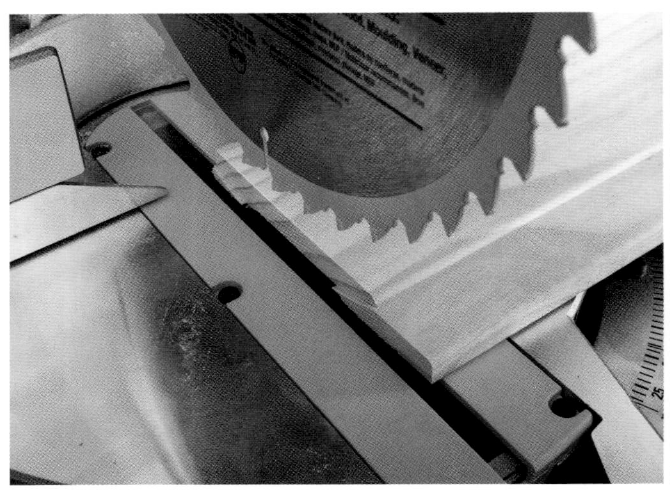

1 **CUT ONE SIDE OF THE PILASTER.** With the piece laid on the miter saw's base, cut the end of just the thicker portion. The cut should be the same dimension as the thickness of the upper, thinner portion of the molding.

2 **CUT THE OTHER SIDE.** Turn the piece over and cut the other side, taking care not to cut the thinner portion so it shows on the face.

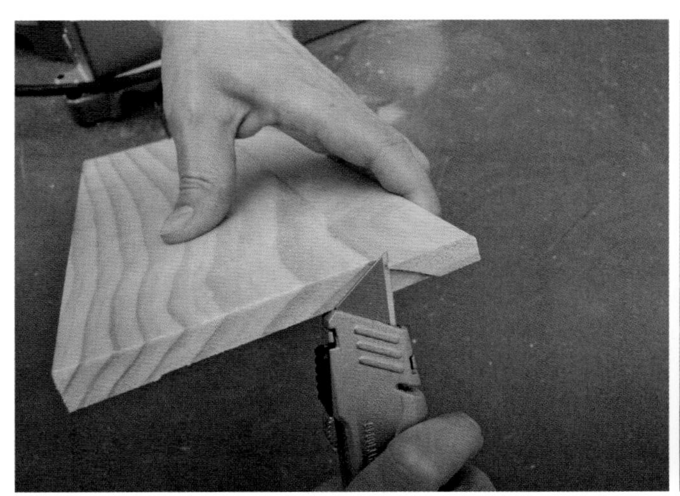

3 **TRIM THE CUT AS NEEDED.** Test that it fits against the butt-cut piece installed on the wall, and trim further as needed.

4 **ATTACH THE PILASTER.** Once you get a tight joint, nail to the wall.

Base Shoe at Door Casing

Where the base shoe meets the casing, mark and cut it at a 60-degree angle, to make an elegant transition.

5 **ATTACH THE BASE CAP.** **The base cap is installed in the standard way. Notch the butt-cut pieces of base cap where they meet at an inside corner (left). Cut the coped pieces as per usual (center), and make any needed micro-adjustments before attaching them (right).**

Base Molding with Corner Blocks

Corner blocks cost a bit more, but they add a distinctive touch to base molding. And there's an added plus: Using blocks, you can cut all the base molding pieces straight, at a simple 90 degrees, saving the trouble of making mitered and coped cuts.

PLINTH BLOCK. Casing with a plinth block at the bottom is a good way to complete an ensemble with base molding and corner blocks.

1 CORNER BLOCK MARKED FOR CUTTING. **Choose corner blocks for inside and outside corners that coordinate with your base molding and shoe. The block shown here is thick enough so the base shoe can dead-end into it gracefully, but it is too tall for this molding, so it will be cut to length. This block can be used as an inside or outside corner.**

Avoid using construction adhesive that dries quickly; you want 10 minutes or more to adjust pieces in case they do not fit tightly right away.

2 CHECK FOR SQUARE TO FLOOR. **Because wall surfaces sometimes curve or wave near the floor, check the wall for square to the floor in both directions. You may need to dig away a bit of drywall to get a block to fit securely.**

3 GLUE THE BLOCK. **Apply a fairly thick bead of construction adhesive to both back sides of an inside corner. The thick bead will allow you to adjust the block slightly for a tight fit against the baseboard.**

1. PREP THE GAP.

2. TAP IN THE NARROW PIECE.

3. SCRIBE FOR NEXT PIECE.

4. ATTACH THE BASE SHOE.

Filling a Small Space

Sometimes door casing is near an adjacent wall, so a very short piece of base molding is needed. Driving a nail into this piece will likely cause it to split. But as long as it fits snugly, a bead of construction adhesive will hold it in place.

1. Install the small piece first, before the longer piece. Measure for the piece's length at both the top and the bottom. You may need to scrape the wall to make room for the piece.

2. Cut a piece to fit snugly, but not too tightly. Apply a dab of construction adhesive to its back, and tap it into place.

3. The small baseboard piece may not be square to the floor. Try tapping its bottom to bring it vertical. If that does not work, scribe the longer piece and cut the reverse miter at that angle, then make the cope cut.

4. If you are installing base shoe, it often looks best to simply have it dead-end into the shorter piece.

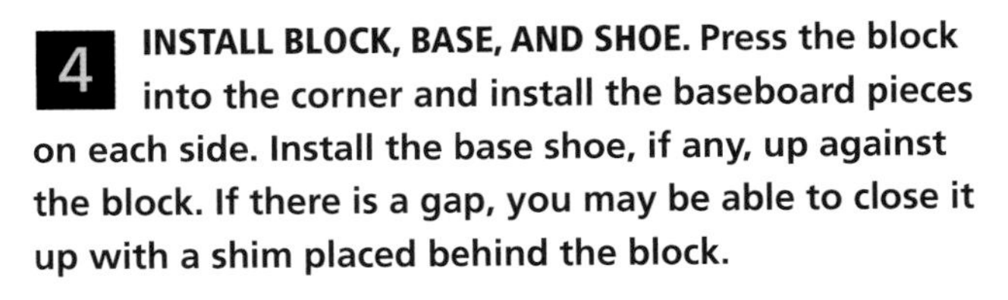

4 INSTALL BLOCK, BASE, AND SHOE. **Press the block into the corner and install the baseboard pieces on each side. Install the base shoe, if any, up against the block. If there is a gap, you may be able to close it up with a shim placed behind the block.**

5 OUTSIDE CORNER BLOCK. **Most blocks can also be installed at outside corners. If there is a gap between block and base, you can usually solve the problem by driving nails through the block and into the base, or into the wall framing.**

1 **TEST THE MITER.** Making a coped joint with picture rail is somewhat difficult because of the rounded top. Cut two scrap pieces at 45-degree miters and hold them against the wall to test the fit. If there is a noticeable gap in front, micro-adjust the miters (p. 89) and retest until you get a neat fit.

Picture Rail

Picture rail originally was designed for hanging artwork: Special metal clips (which can still be purchased) grab onto its top rounded edge. Wire or string reaches down from the clips to suspend framed pictures. This old-fashioned use is making a comeback, but nowadays it is more common to install picture rail for decorative purposes only.

In some homes the gap between the top of the picture rail and the ceiling gets filled with caulk. In other homes picture rail is installed tight against the ceiling. For a more authentic and nuanced effect, install picture rail with a visible gap, about 3/8 in. down from the ceiling.

TIP **When picture rail is not installed up against the ceiling, it may not cover imperfections in the wall-and-ceiling corner. If you need to hide cracks, crown or cove molding may be a better choice (see Chapter 8).**

2 **SHIM AND NAIL.** Cut pieces to fit, and test the fit. Near a corner, place the thick end of a shim (which is 3/8 in. thick) against the ceiling, push the rail up against it, and drive nails to attach.

3 **USE A STRAIGHTEDGE.** If the ceiling is somewhat wavy, place a straightedge under the molding to maintain a fairly straight line. Have a helper watch as you position the molding and determine the best-looking arrangement. You may choose to "split the difference" between straightness and following the curves of the ceiling.

Plate Rail

Plate rail can supply a quaint cottage-style touch to a kitchen or dining room. The plain casing shown here has several grooves in its back that work surprisingly well for keeping plates from sliding.

Plate rail is typically installed about 60 in. above the floor, but you may want to install it higher if you have 10-ft. ceilings or lower if it is in a low-traffic area.

PLATE RAIL WITH UPSIDE-DOWN CASING. **A piece of plain oak casing, sometimes called sanitary casing, has milling grooves on its back side and the front looks like plain shelving. This design uses two casing boards and one small crown molding board. Nail the back edge of the shelf to the top edge of the other casing board. Nail the crown piece to complete the shelf. Make long completed sections of this ensemble, and cut them to fit on the wall.**

GROOVED PLATE RAIL. **To make a more elaborate plate rail, start with a 1× shelf that is 5 in. or wider. Use a router with a grooving bit—either on a router table or with a guide—to cut a groove about 2½ in. from the back of the shelf. Attach the shelf, groove-side up, to a base piece, and add crown molding (as shown) or evenly spaced decorative brackets to hold the shelf level.**

Chair Rail

Chair rail can be a pleasant decorative element in the middle of the wall. It is usually installed about 3 ft. above the floor, and most often runs around an entire room. Traditional chair rail has a curved profile, but modern and Craftsman rails may be made of plain 1× boards. Chair rail or rail cap is often installed at the top of wainscoting (see Chapter 9).

CHAPTER SIX

DOOR AND WINDOW JAMBS AND STOOLS

BEFORE YOU CAN INSTALL door and window casing—which lies flat against the wall—a window or door needs to have a jamb, laid against the framing, with its edge flush to the wall surface. If your windows or doors already have jambs installed, you can probably skip this chapter.

Although this book is not about installing doors or windows, this chapter will briefly show how to install a pre-hung door. In addition it shows how to install window jamb extensions, which are often needed after a new window is installed, or after remodeling has changed the wall surface.

Jambs serve practical as well as decorative functions. For instance, one side of a door jamb holds the door's hinges, and the other side houses the strike plate for the door's latch. Consequently, jambs need to be installed more firmly than most trim pieces.

This chapter also covers window stools, the horizontal ledges at the bottom of a window on the inside. (Note that what most people refer to as a "sill" is more correctly called a "stool"; the word "sill" actually means the outside piece.) Stools can be a bit tricky, but with careful planning and attention to detail they are not difficult to cut and install.

Installing a Door Jamb

If you are replacing an interior door, you could leave the existing jamb in place and just replace the door and casing, but chances are the old jamb is banged up and out of square. In most cases, removing the jamb and installing a pre-hung door (which comes with a jamb already cut and exactly sized to fit) will be the most efficient use of time and will lead to the neatest looking job.

See pp. 52–53 for general tips on checking an opening for square and plumb, and attaching shims to straighten things out.

TIP **If a door is very heavy, you may choose to remove it from the hinges, install the jamb, and then rehang the door. But for most interior doors it is easiest to install with the door attached to the jamb. That way, you can more easily check that the gap between door and jamb is a consistent width all the way around as you work.**

1 **CHECK THE OPENING.** Measure the rough opening to be sure the jamb will fit. If the opening is out of square or not plumb, set the door in place and check with a level to be sure it will fit after it has been plumbed. If not, you may need to cut out a stud and widen the opening.

2 **CUT THE BOTTOM OF THE JAMBS.** The jamb sides will likely be longer than you want them to be, so cut them to the desired length. Draw a square line and cut with a circular saw.

Straighten a Twisted Stud

Especially on the door's hinge side, it is important that the jamb not be twisted—that it be perfectly perpendicular to the wall surface.

1. Check that the stud is square to the wall.
2. If not, install tapered shims as needed to correct the situation.

1. CHECK FOR TWIST.

2. SHIM TO FIX.

TIP **HOW HIGH SHOULD A DOOR'S BOTTOM BE ABOVE THE FLOOR?** This is a matter of debate among carpenters. Some allow a gap of as much as an inch between floor and door—which will look fine, and will allow for future installation of new flooring. But most recommend a height of 3/4 in. to 1 in. under a bathroom door, to allow for ventilation; elsewhere, go for 1/2 in. to 5/8 in. If the floor is carpeted, take into account the thickness of the carpet's pile and have the door 1/2 in. or so above that.

3 **SHIM THE HINGE SIDE AS NEEDED.** If the stud on the hinge side is perfectly plumb, you can simply fasten the jamb to it. But studs are rarely perfect, and shims are usually needed. Measure to find the heights of the hinges (there may be two or three), and attach shims, if needed, at those heights. At each spot nail two tapered shims, one in each direction, so the jamb will not get twisted.

TIP Check that the jamb will be flush with the wall on both sides of the doorway. If the jamb is only slightly narrow (about 1⁄8 in., say), you can split the difference; casing can cover a 1⁄16-in. discrepancy. If the jamb is too wide, install one side flush and plane the other side after you have installed the door.

4 **CHECK THE DOOR AGAIN.** Place the door in the opening, with the hinge side of the jamb pressed against the shims you just installed. Check again to be sure you will be able to install the door with its top level, the other side plumb, and with a consistent gap (usually 3⁄32 in. or 1⁄8 in.) between door and jamb all around.

5 **JAMB FLUSH WITH WALL?** Use a straightedge to check that the jamb's edge is flush with the wall surface. If it is not, the casing will not install properly.

Shimming a Jamb Bottom

Because the world is not perfect, you may need to shim one of the jamb's sides up from the floor in order to keep the jamb's top level. Shimming up a small distance, like that shown at left, is fine. But if you need to raise it more than 1⁄4 in., it usually looks best to cut the other jamb instead.

TIP **SHIM AND NAIL THE TOP JAMB?** Some carpenters install a set of shims in the middle of the top jamb piece and nail it in place. However, for a door that is less than 36 in. wide this is not usually considered necessary; the nails used when installing casing will keep the top jamb from sagging.

6 **NAIL THE HINGE-SIDE JAMB.** Holding the jamb flush with the wall surfaces, drive nails through the hinge-side jamb near the hinges.

7 **SHIM AND NAIL THE LATCH-SIDE JAMB.** Attach shims to the other (latch) side, to maintain an even gap between door and jamb. Drive nails through the latch-side jamb at the shim locations.

8 **DRIVE ONE LONG SCREW.** Once you are happy with the door installation, remove one screw from each of the hinges. Replace it with a long screw that penetrates at least 1 in. into the wall stud. (Alternatively, remove all hinge screws, open the hinge, and drive a hidden screw as shown in the next step.)

9 **SCREW THROUGH THE LATCH LOCATION.** Slip in a pair of shims near the strike plate. If you have a long screw with a head that matches the strike-plate screws, use the method shown in Step 8 and install it as a replacement for one of the strike-plate screws. Or remove the strike plate (if it came installed on the jamb), drill a pilot-and-countersink hole so the screw's head will not be proud of the wood surface, and drive a long screw at least 1 in. into the wall stud.

10 **TRIM THE SHIMS.** Cut the shims flush with the jamb edge and wall surface. Some carpenters use a hand saw for this. Others prefer to slice a couple of times with a sharp knife, then break the shim off; that method ensures against marring the wall and the jamb.

WINDOW ASSEMBLY

When a window's frame does not extend out to come flush with the wall surface, you need to build a jamb (or jamb extension) to cover the wall framing and to provide a surface for attaching the casing.

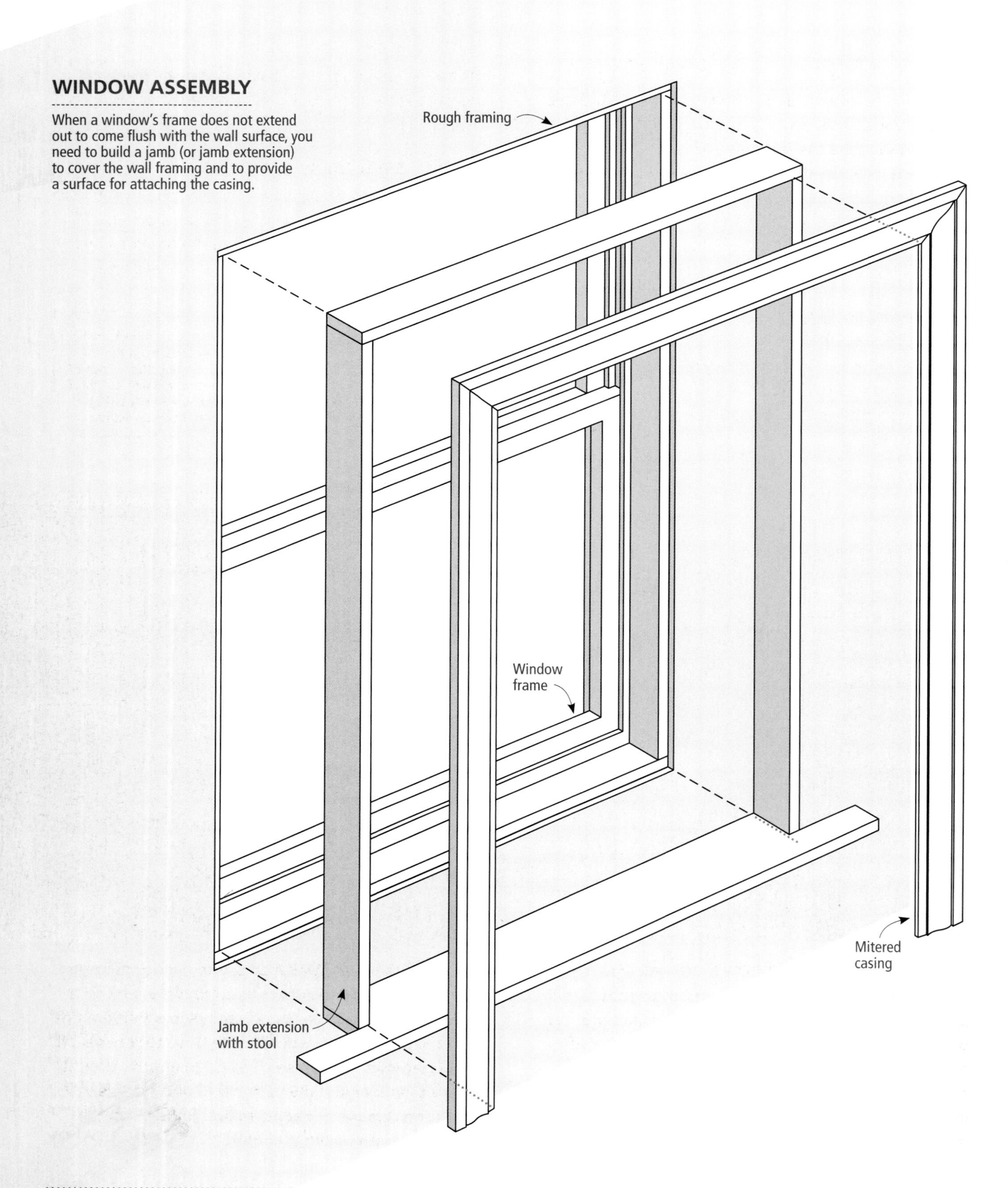

Installing a Window Jamb

A window comes with an outside frame along which the window sashes slide. (On a casement window that cranks open and shut, the frame houses the sash's hinges and latch.) In some cases this frame extends inward far enough that its front edge can be flush with the wall surface: In that case, it also acts as the jamb. But often the frame comes well shy of the wall, and in that case you need to install a jamb, or "jamb extension."

Some windows come with jamb extensions. An extension may be adjustable, or you may need to rip-cut it to the needed width. Often, however, it's up to you to make the jamb. A window jamb is usually made of the same material as the window and the casing, but if the window and casing are of different materials, the jamb can match either element.

As can be seen on p. 112, the stool may be made at the same time as the jamb, or it can be added after the jamb is installed.

Prepping the window and wall

If the window is poorly installed, or if the surrounding wall is uneven, installing a jamb will be difficult. If possible, adjust the window for a better fit and straighten the wall. That will make the trimming easier, and will lead to a neater appearance.

TIP **You'll notice that the new window shown on the following pages is installed into an opening that is out of square and not level and plumb. This is not unusual in an older home. As long as the window itself is square, you should be able to cover up the imperfect opening with little difficulty.**

1 CHECK THAT THE WINDOW IS SQUARE. If it is not, trimming will be difficult and the window will probably not operate smoothly. If the window is out of square, look carefully for mounting screws and remove at least some of them. You may also have to remove some outside trim, if that was previously installed. Pry the window into square, and drive new mounting screws.

2 MEASURE FOR JAMB WIDTH. The window should be installed so it is recessed from the wall by a consistent distance. That way, jamb pieces that are rip-cut to the same width will all come flush to the wall. If it is not recessed evenly, see if you can remove mounting screws and adjust the window in or out as needed. If you cannot make the adjustment, you will need to rip-cut the jambs at slight angles.

3 **LEVEL THE WALL SURFACE.** If the wall is noticeably wavy or uneven, you can apply layers of joint compound to even things out. Use fast-setting compound and apply with a wide taping knife. It may take several coats, with scraping and sanding between, to get the wall reasonably close to straight.

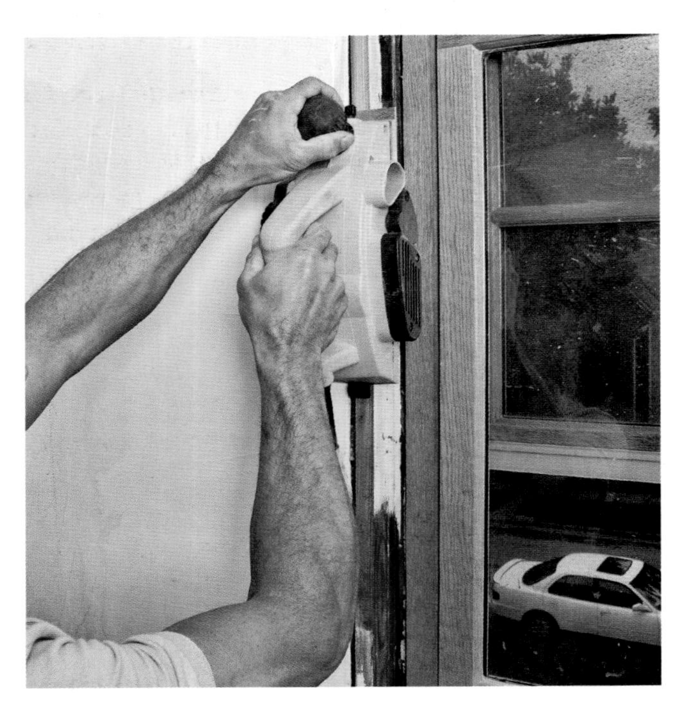

4 **PLANE THE FRAME EDGES.** If any wood parts extend proud of the wall surface, plane them down so they are flush with the wall. (In this case, the new window was installed inside an older frame, which needed planing.)

Simple Narrow Jamb Extension

If the window frame is just an inch or two short of the wall, simply cut pieces for a jamb extension and nail them to the window. Do not install the pieces flush with the inside edges of the frame; instead, maintain a consistent reveal of 1/4 in. or more, as shown below.

TIP **Instead of trying to install a jamb so that its inside edges are flush with the inside edges of the window frame, the inside dimensions of the jamb box should be a bit wider and taller, to produce a consistent "reveal" all around the window's frame. For a vinyl window this reveal may be an inch or more. For a wood window, like the one shown here, the reveal is typically 3/16 in. or 1/4 in.**

Building a basic jamb

The simplest jamb is made of 1× boards rip-cut to fit and assembled at the corners with screws. After shimming, the jamb attaches to the opening's framing with face nails.

TIP **If the window will receive picture-frame casing (pp. 138–140), build a four-sided jamb. If you will be installing a stool (pp. 110–112), make the jamb with three sides, and attach the stool to its bottom.**

1 **MEASURE FOR THE JAMB PIECES.** **First measure for the jamb's depth (the width of the pieces)—the distance between the window frame and the wall surface. Then measure the width of the window (above) and the height (inset). Keep in mind the reveal (see the tip on the facing page).**

2 **RIP THE JAMB PIECES.** **Use a tablesaw (or a circular saw with a rip guide) to rip-cut jamb pieces. Cut the pieces to length to produce a jamb that is slightly larger than the window frame and creates a consistent reveal all around. See the tip on p. 108 for measuring the lengths.**

3 **ASSEMBLE THE JAMB.** **Place the three (or four, if you will have picture-frame casing with no sill) pieces on a flat surface, position the pieces, and double-check that the resulting jamb ensemble will fit in the opening and produce a reveal on the window frame. Attach simply by driving nails or by drilling pilot holes and driving screws.**

If the window will have a stool, cut and attach it following the instructions on pp. 110–112. If it will have no sill and will get picture-frame casing, add a fourth piece of jamb at the bottom, the same length as the top piece.

Rabbetted jambs

The simple jamb shown in the steps on pp. 105–107 has two disadvantages: You end up with exposed nails, which must be sunk, filled, and sanded; and the jamb is not actually attached to the window frame, so it can come apart slightly over time.

For these reasons, some windows come with rabbeted jambs that fit into the window frame, and have holes for driving screws to attach the jamb with no visible fasteners. If your window does not have these, you can make your own rabbeted jambs, as shown in the following steps. (For another option, consider attaching the jamb with pocket screws, as shown on the facing page.)

TIP **Assuming you are building with ¾-in.-thick lumber, the top piece should be the total width of the window frame plus 1½ in. (because it will span the thickness of the two side jamb pieces), plus two times the desired reveal. The side pieces should be the total height of the window frame minus the thickness of the stool, plus two times the desired reveal.**

1 **MARK THE RABBET AT THE END OF THE BOARD.** **Cut the jamb pieces to width and length. Determine which side of each board you want to show and, to keep yourself from getting confused, roughly draw the outline of the angled rabbet you will cut. The side and the edge that is not rabbeted will be on display when the jamb is exposed.**

2 **MARK FOR SCREW LENGTH.** **Plan to drive long screws through the boards; the wider the jamb, the longer the screws. Mark where the screw head will be when the screw's tip pierces the jamb by about ½ in. (inset). Adjust a tablesaw's blade to a 4-degree angle, and raise it to the mark you just made (above).**

3 **CUT THE RABBET.** **Adjust the saw's fence to make a cut that is about a blade's thickness inside the side of the jamb board. Hold the board firmly against the fence as you make the long tapered cut (above). After you have cut all the jamb pieces this way, readjust the blade to make a 90-degree cut that is just high enough to meet the top of the tapered cut. Make the cut (inset) and remove the waste.**

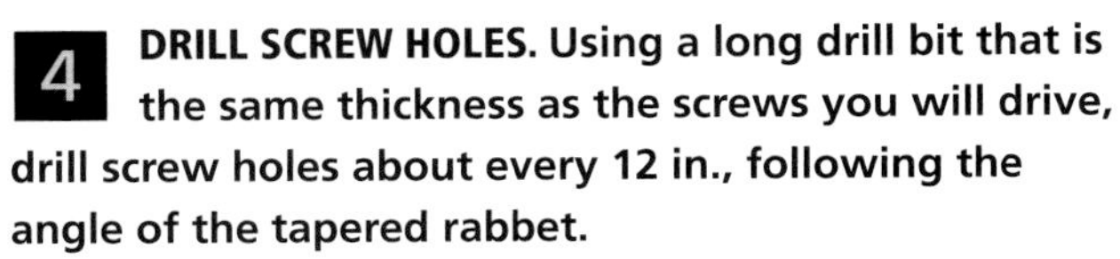

4 **DRILL SCREW HOLES.** Using a long drill bit that is the same thickness as the screws you will drive, drill screw holes about every 12 in., following the angle of the tapered rabbet.

5 **SET THE SCREWS.** Set the jambs in place and start the screws in their holes. Clamp the jamb precisely in place, maintaining the desired reveal, and drive the screws into the window's frame.

Pocket-Screw Jamb

You can also attach jambs using pocket screws. (However, this method requires that you have ½ in. or more of space to the sides of the jamb; if the fit is tighter, rabbeted jambs work better.)

1. DRILL POCKET-SCREW HOLES.

2. DRIVE THE SCREWS.

1. Cut the jamb pieces to width and length and attach a stool or bottom jamb piece. Use a pocket-screw jig to drill pocket holes every 12 in. or so. Assemble the jambs.

2. For attaching to a hardwood window frame, use fine-thread screws; for softwood, use coarse-thread screws. Start the screws in the holes. Clamp the jamb in place with a consistent reveal all around, and drive pocket screws into the window frame.

Making and Installing a Stool

A window stool makes a horizontal shelf that can be simply decorative or useful as well. It must be planned with the casing in mind. For a harmonious appearance, the stool's horns (the narrow portions at each end) should run past the casing evenly on both sides. See the drawing below for determining the length and width, as well as the size of the horns.

TIP **If you want to set flower pots or knickknacks on the stool, make it 5 in. wide or wider. However, if doing so will cause it to overhang the apron by more than 2 in., it's a good idea to add a piece of cove or other molding to the top of the apron, to provide more support for the stool.**

1 MARK THE WALL FOR THE SIDE OF THE CASING. If you're installing a stool, set the three-sided jamb assembly in place next to the window to help you measure. Position both jamb sides to produce the correct reveal on the window frame. Hold a piece of the casing you will use (see next chapter) as shown, against the wall, with the desired reveal on the jamb. Mark the wall to indicate where the side of the casing will be. Do the same on the other side of the window.

WINDOW STOOL DIMENSIONS

A stool's overall length, including the horns, is the distance between the outside edges of the casing, plus twice the desired underhang. Stool width can be as much as 2½ in. wider than the width of the jamb.

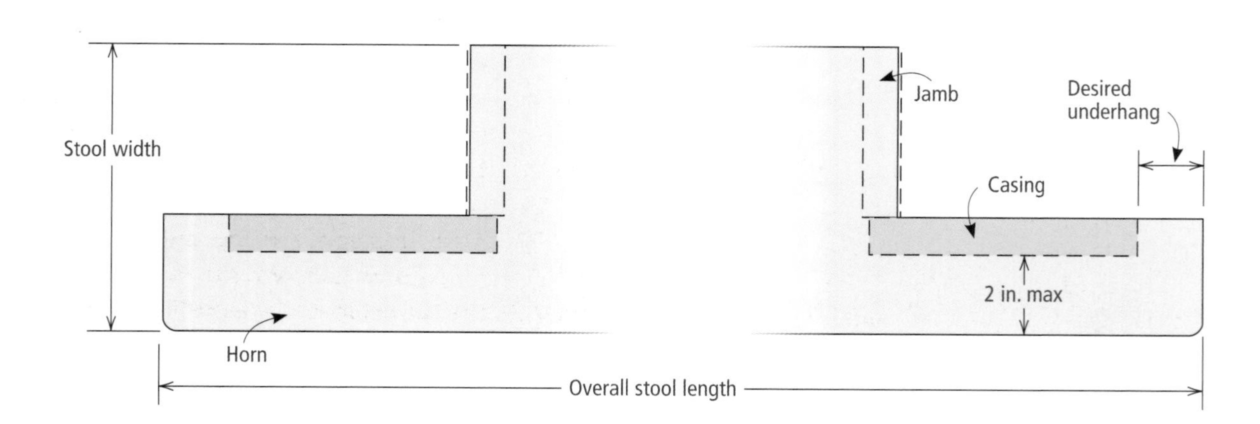

2 **MEASURE FOR THE LENGTH OF THE STOOL.** Measure the distance between the two marks on the wall; this is the distance between the outside edges of the casing. In most cases, the stool will be longer than this, so it can underhang (run past) the casing on each side. In our example, the underhang will be 1 in.

3 **MEASURE FOR THE DEPTH OF THE HORN CUT.** The depth of the cut should be the same as the width of the jambs. Check with a straightedge held across the wall and the edge of the jamb to be sure.

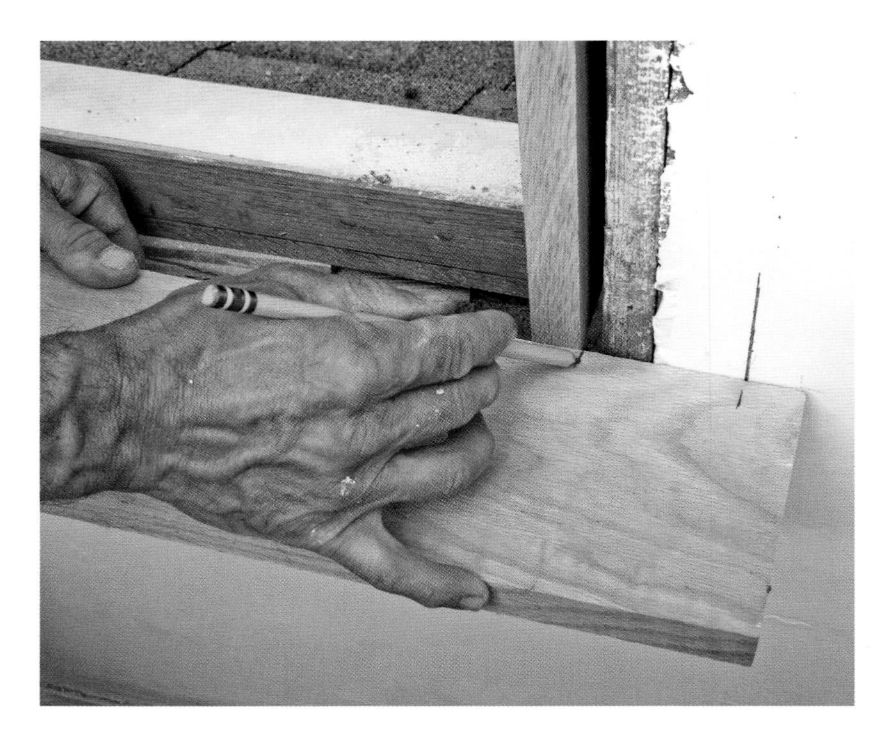

TIP It is typical for stools to run past casings by 1 in. to 1½ in. But you may choose to match existing underhangs on other windows in the house, for consistency.

4 **MEASURE FOR THE LENGTH OF THE HORN CUT.** Cut the stool to the desired overall width (see the illustration on the facing page). Make a mark showing the distance it will run past the casing, and hold it against the wall with the mark aligned with the casing mark you made in Step 1. Position the jamb where it needs to be in order to maintain the correct reveal against the window frame, and mark for the horn cutout on the outside of the jamb's thickness.

5 **CUT OUT THE HORN.** Transfer the measurements onto the end of the board. Cut first with a circular saw, chopsaw, or tablesaw, then complete the cut with a jigsaw or handsaw.

6 **ROUND OVER THE EDGES.** Equip a router with a self-guiding roundover bit and test on scrap pieces until you get the desired roundover radius. Round over the stool's front edge and the sides of the horn.

7 **ATTACH THE STOOL.** Attach the stool to the underside of the jamb assembly (from p. 107) with a pair of screws or 2½-in. finish nails at each joint.

8 **ATTACH JAMB AND STOOL.** Set the jamb-and-stool assembly in place and test for a good fit all around. Install shims under the stool as needed. Remove the assembly and apply a thick bead of construction adhesive for the stool to sit in. Set the assembly in place and fasten with screws or nails (depending on whether the jamb is plain, rabbeted, or pocket screwed).

Trimming a Bay Window

If you buy a bay window that comes assembled as a single unit, trimming it is simple. In fact, it may come with all or most of the trim you need. But in an older home with a bay that is made of three separate windows, things can get more complicated.

The steps on the following pages can help you address a number of real-life situations, where you may need to trim areas with odd and inconsistent angles and lines that are not quite parallel. The process involves plenty of marking boards in place, using scrap pieces to figure angles, and initially cutting boards a bit long, then "sneaking up" on the final cut by shaving off a little at a time. You may also use a chisel or saw to modify the walls' framing in order to make trim boards lie flat.

When trimming a bay window, first install three stools with odd-angled miters where they meet. Then rip-cut and install jamb extensions that have front edges that match the plane of the walls, so your casing can sit flat against the walls. Then the casing itself must be cut at bevels with angles that match the angle of the windows and walls.

1 **CAPTURE THE STOOL ANGLE. Especially in an older home, the angle of the windows can be odd, and may not be repeated exactly on each side of the middle window. Cut a notch in a couple of scrap (but straight) boards so they can rest against the windows. Hold them in place and capture the angle with a T bevel (left). Trace this angle onto a board (right).**

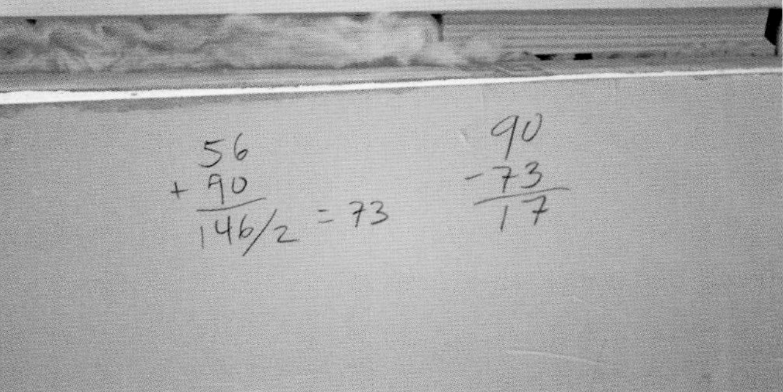

2 **FIND THE ANGLE. Hold a Speed square against the traced line to find its angle (which in this case is 56 degrees) (left). Because a miter saw configures a 90-degree angle at 0 degrees, and your cuts for the stool will be half the desired angle, do some simple math (right): Add 90 to the angle of the traced line and divide by 2. (In our example, 56 + 90 = 146; 146 divided by 2 = 73. Then subtract that number from 90 to get the angle at which you need to cut the stool pieces.)**

3 **MAKE THE STOOL RETURN.** Cut the ends of the board you will use for the stool at 45 degrees and attach the cutoff pieces to make a return, so the ends will have the same profile as the front (left). Place a scrap of the casing you will install on top of the stool to determine how much you want the stool to run past it (right).

4 **MEASURE THE DEPTH OF HORN CUT.** Hold a scrap piece of wood against the wall so it follows the wall's line in the way the casing will when it is installed. Use a sliding square to measure the distance from the window to the inside edge of the scrap piece. This is the depth of the cut for the stool's horn (left). In our case, the two outside windows are at different depths; this is not unusual, so be sure to measure horn depths at each window (right). If the stool needs to be notch-cut to match the window, use a sliding square to measure for this as well (inset).

5 **MEASURE THE LENGTH OF HORN CUT.** Hold a piece of casing (in this case, the casing also includes backband) in place and mark the wall at its outside edge (left). Hold a scrap piece of stool overhanging the casing the amount you want, mark that (right), and measure for the length of the horn (inset, showing measuring the other window).

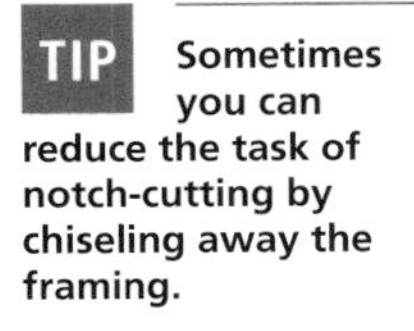

TIP Sometimes you can reduce the task of notch-cutting by chiseling away the framing.

6 **MARK THE HORN CUTS.** Use a sliding square to draw the basic rectangle for the end cuts (left). Then use it to draw the notch cut. Be sure to erase or cross out the outer corner of the rectangle, so only cutlines remain (right).

7 **HOLD IN PLACE AND MARK.** Where the stool pieces meet at angles, marking and cutting notches can get a bit hairy. Where possible, leave the boards long as you make the notches, then cut to length at the end. Hold each piece against the framing and mark for the notches, using the same techniques as for the end notches.

8 **CUT THE NOTCHES.** Make rip cuts using a tablesaw, and crosscuts with a jigsaw. Cut the board a bit long, so you can sneak up on the cuts. Take your time.

TIP This project, which shows trimming a bay window, mostly deals with jambs and stools—the hardest part of the job. However, some casing work is also shown. For more information on casing, see Chapter 7.

9 **MARK THE STOOL FOR CUTTING.** Once you've got the notch cuts so they fit against the window and the framing, lay the pieces on top of each other. Carefully mark the place where they meet in front. Cut the pieces ⅛ in. or so too long, set them back in place, and mark for the final cut, making sure your angle is correct, to produce a neat joint.

10 **MEASURE FOR AN ANGLED CUT AT THE HORN.** Where the wall is angled, the horn may not meet it neatly. Measure for an angle cut, and cut with a tablesaw, not using the fence.

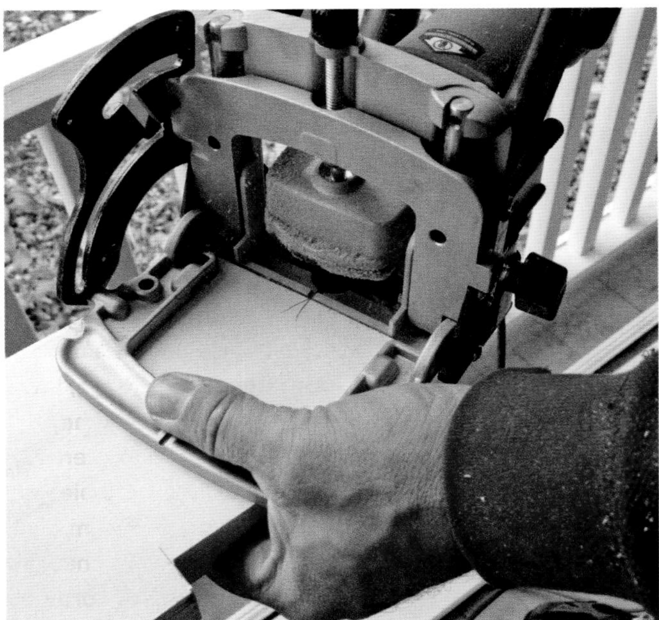

11 **BISCUIT-JOIN THE STOOL.** Once you are happy with your cuts and the joints are nice and tight, use a biscuit joiner to attach the stool pieces where they meet at angles.

12 **INSTALL JAMB EXTENSIONS.** To prepare for jamb extensions, cut away drywall or plaster as needed (left). Cut jamb extensions to the correct width so they are flush with the wall surfaces, and nail them in place (right). For more about jamb extensions, see pp. 106–109.

13 **MEASURE THE TOP CASING ANGLE.** The top casing pieces need to be bevel cut at the ends where they meet at angles, and the vertical casing pieces need to be rip-cut at bevels to match the angles as well. Measure for the angle the same way you did for the stool, in Steps 1 and 2.

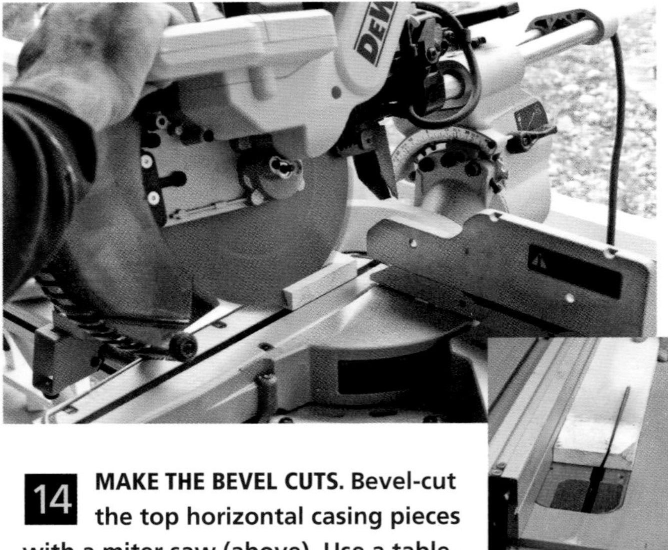

14 **MAKE THE BEVEL CUTS.** Bevel-cut the top horizontal casing pieces with a miter saw (above). Use a tablesaw to make rip bevel cuts for the vertical casing pieces on each side of the front window (inset).

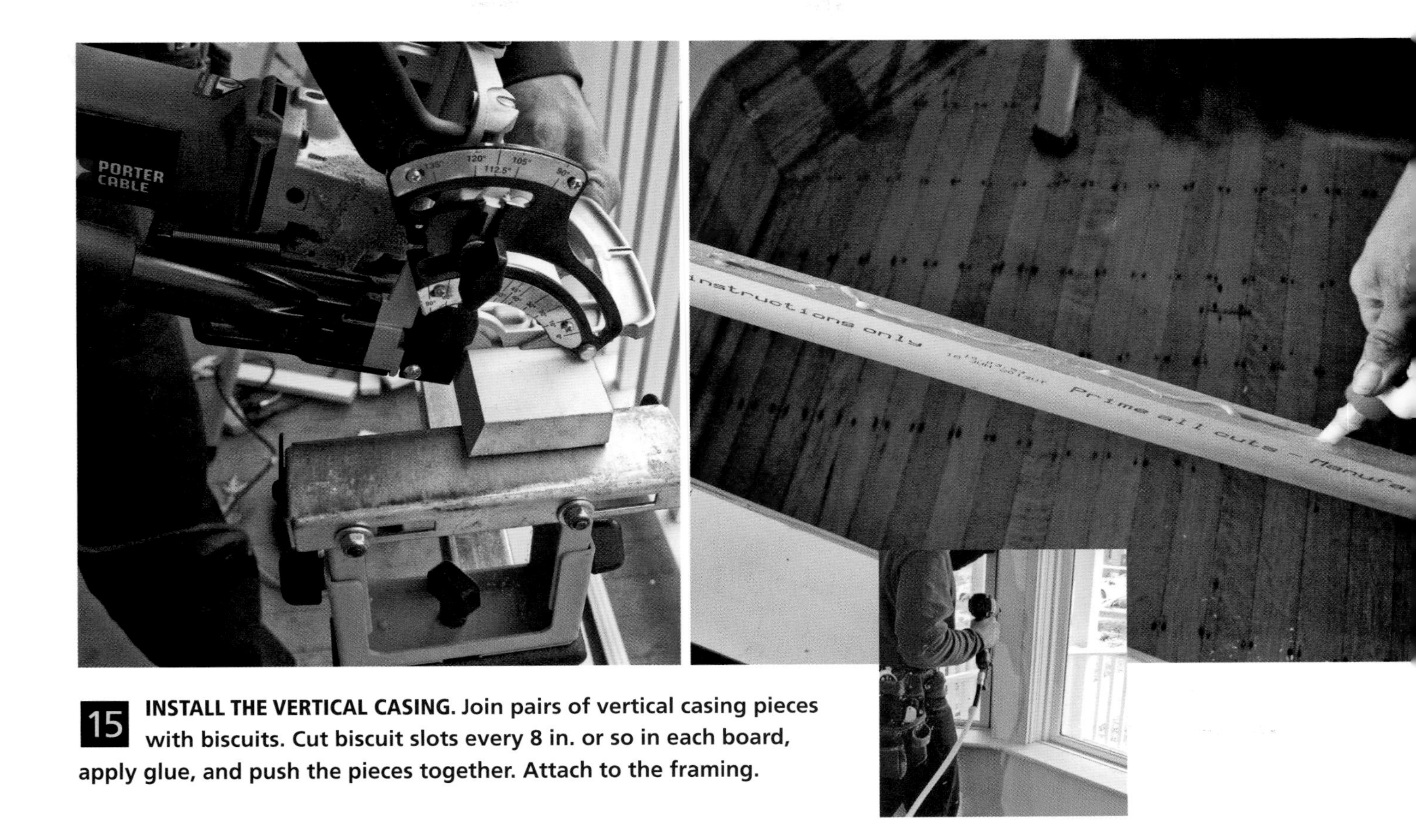

15 **INSTALL THE VERTICAL CASING.** Join pairs of vertical casing pieces with biscuits. Cut biscuit slots every 8 in. or so in each board, apply glue, and push the pieces together. Attach to the framing.

16 **FINISH THE BAY WINDOW.** Take a day off from all this hard work, then carefully caulk, spackle, and paint.

CHAPTER SEVEN

DOOR AND WINDOW CASING

ONCE DOORS AND WINDOWS are fitted with jambs, it's time to apply the most visible part of trim, the casing. This chapter covers casing for both windows and doors, because installation procedures are much the same for each. However, there are differences: Windows have a bottom horizontal piece—either an apron under the stool or a bottom piece of casing for picture-frame molding. And door casing abuts baseboard trim at the bottom.

Casing may be a simple one-piece affair, as with mitered or Craftsman-style trim. Or it may be an ensemble of three or more trim pieces that work together to produce a rich appearance. This chapter shows a good number of possibilities. Using the techniques shown, you can easily produce a grouping of your own design.

You may need to match existing molding in the room or in nearby rooms. If your home is fairly new, you may be able to find trim pieces that match exactly. An older home may have trim profiles that are no longer manufactured. In that case, you can either pay for a mill to make exact copies, use trim pieces (perhaps several in combination) that come pretty close to mimicking the existing trim, or produce similar-looking trim pieces yourself using a router.

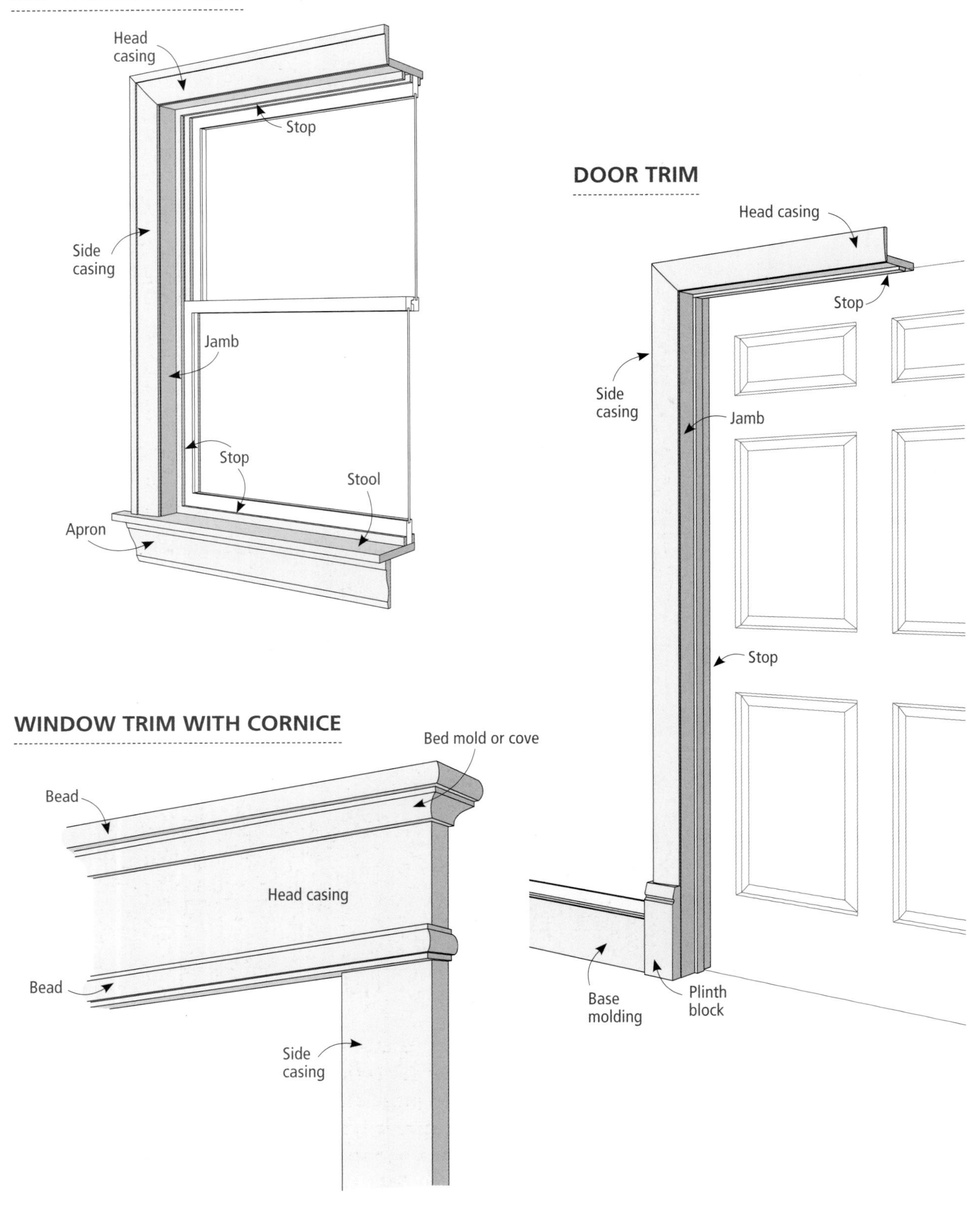
BASIC WINDOW TRIM
Head casing
Stop
Side casing
Jamb
Stop
Stool
Apron
DOOR TRIM
Head casing
Stop
Side casing
Jamb
Stop
WINDOW TRIM WITH CORNICE
Bed mold or cove
Bead
Head casing
Bead
Side casing
Base molding
Plinth block

Installing Mitered Casing

Mitered casing is usually composed of single pieces cut at 45-degree miters to produce neat 90-degree corners. For a richer appearance, other elements such as backband can be added (p. 129). For an overview of casing types, see pp. 21–23.

Though mitered casing is straightforward to install, it's easy to make mistakes and end up with sloppy joints. The instructions here walk you through the process with what may seem like painstaking detail. But a little extra time spent on installation will help you avoid costly and maddening mistakes.

Before you begin, check that the jamb edges are flush with wall surfaces on both sides of the opening. If there is a serious problem—more than ⅛ in.—it may be worth your while to detach and reattach the jamb. If the wall is a bit proud of the jamb edge, you may be able to solve the problem by tapping or pounding with a hammer and scrap piece of lumber, as shown at right. If the jamb is proud, you may be able to plane it down, or build out the wall with joint compound (see p. 106).

CHECK THAT JAMB AND WALL ARE FLUSH. If the wall is slightly proud of the jamb edge, you may be able to pound the wall surface flush with the edge of the jamb.

1 CHECK FOR SQUARE. Use a framing square or large Speed square to check the jamb corners for square. If they are not, you may be able to remove shims, pry, and add new shims to bring the jambs square. If not, you will need to micro-adjust your miter cuts on the casing pieces.

2 CHECK WITH SCRAP PIECES. Also check that your miter saw is accurate. Cut two scrap pieces at 45 degrees and hold them together against the jamb. If the jamb is square but the two pieces do not follow the jamb, take the time to precisely adjust your saw (see p. 56).

1. MARKING THE REVEAL WITH A SQUARE

1A. CUTTING THE RABBETS

1B. THE FINISHED REVEAL GAUGE

2. USING THE REVEAL GAUGE

Marking Reveals

Every style of casing, from the simplest to the most complex, is installed with a reveal on the jamb edge. Jambs are usually only ½ in. thick, though some are ¾ in. Because you need room to nail the casing to the jamb, the reveal is usually just 3/16 in. or ¼ in.

To mark for a reveal of consistent width, you can use a sliding square (top left) or a reveal gauge. To make a reveal gauge:

1. Start with a piece of ½-in. or ¾-in. plywood about 4 in. square. Set a tablesaw's blade to cut at a depth of about ¼ in. Clamp a board to the edge of the rip fence (because the blade should not touch the fence itself), and cut two or more passes in two directions (**1A**) to produce a rabbet that is just shy of the desired reveal (**1B**).

2. You can use a reveal gauge (or a sliding square) to draw a line all along the edge of the jamb, but that would make a line that might show through stain and possibly even paint. Instead, draw light lines at certain places only, or hold the gauge against the jamb and the casing while you drive each nail.

TIP When measuring between the two reveal lines, you may find it easier to "burn an inch," as shown on p. 55.

3 **MARK REVEALS.** Use a sliding square or a reveal gauge (see the sidebar on the facing page) to mark the desired reveal on the jamb edges at the corners. If you're going to paint the molding with paint that can cover pencil marks, you can slide the guide and pencil all along the edges to mark for the reveal all around.

4 **MEASURE FOR THE HEAD CASING.** At the lintel (top of the jamb), measure the distance between the reveal marks. Alternately, measure between the inside edges of the jamb and add twice the desired reveal.

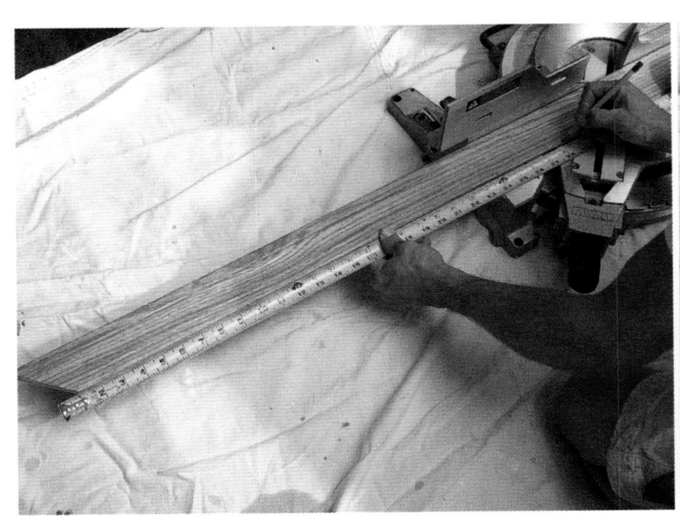

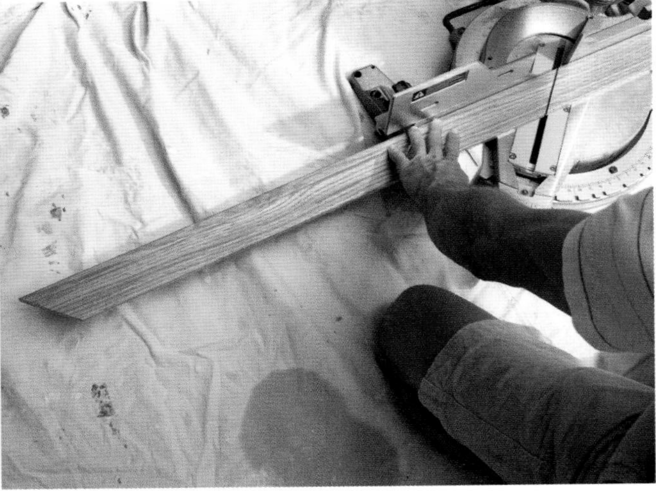

5 **MARK AND CUT.** Cut one end of a casing board at a 45-degree miter. To measure for the other cut, hold the tape measure as shown at left so you "burn an inch." Use the methods shown on p. 58 to cut a 45-degree miter (right).

Three Fixes for Open Joints

If your joints are not as tight as you would like, here are three possible fixes. If the wall is flat but the joint is open at the top or bottom, adjust the miter angle slightly—often less than a degree—on one or both of the boards (1). (Changing the angle on only one piece will lead to its miter being slightly longer or shorter than the other piece, but if you change the angle very slightly, this will not be noticeable.)

Sometimes the problem is due to an uneven wall surface. If there is a gap behind the casing, slip in a shim behind one or both of the pieces (2). If the wall bumps outward, you can either scrape the wall or use a Surform tool or plane to reduce the thickness of the casing (3).

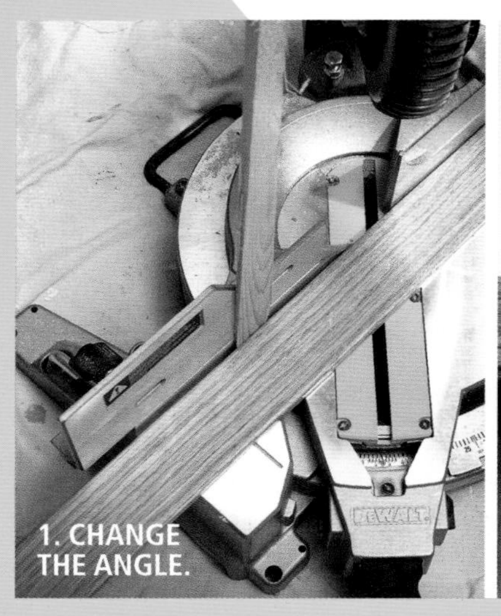

1. CHANGE THE ANGLE.

2. SHIM BEHIND THE CASING JOINT.

3. SHAVE THE BACK OF THE CASING.

TIP Some people use baseboard trim as casing, especially if there is a wide gap to cover. However, because baseboard has a squared-off thick edge, it tends to look out of place on a wall; true casing has a thick but rounded edge for a handsome and finished appearance.

6 **CLAMP THE TOP PIECE.** For a door, clamp the head casing (the top piece) in place, aligned with the reveal lines on each side. Don't drive nails yet. If you are working on a window, tack (partially drive, so you can easily remove them) two or more hand nails. This will allow you to make tiny adjustments to the joints, if needed, when you install the sides.

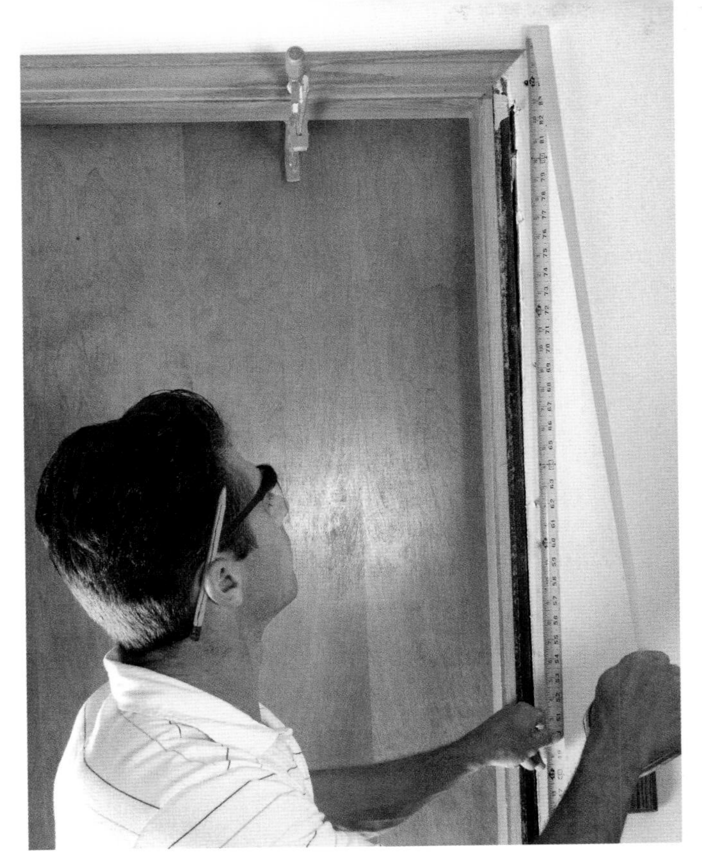

7 **MEASURE FOR THE SIDES.** For each side piece, measure up from the floor (or a plinth block) to the top of the clamped casing. Cut pieces to fit.

TIP Because casing is often not flat, it can be a bit difficult to make a miter cut that ends precisely at the cutline. It is often best to cut a board ¼ in. or so long, then "sneak up" on the cutline with successive cuts until you achieve perfection.

8 **TEST THE FIT.** To ensure a snug fit, cut the side pieces about ⅛ in. long. Test the fit, and shave off the bottoms as needed. Look for a tight fit at each mitered joint. You may be able to achieve perfection by slightly adjusting the head casing.

9 **NAIL THE TOP PIECE.** Once you are sure of the fit, drive several nails to secure the head casing.

TIP Some carpenters prefer to reinforce the corners (Step 13) first, then drive the nails into studs and jambs.

10 **GLUE THE ENDS.** Many carpenters skip this step, but it can add strength to the joint and certainly won't hurt. Apply a bead of wood glue to the cut end of each side piece just before attaching it.

11 **NAIL THE SIDE CASING.** Hold a side piece in place, tight against the head casing, and drive nails to attach the casing. Start about 1 ft. from the top and work your way down, driving a pair of nails—one into the wall stud and one into the jamb—every 16 in. or so.

TIP **NAIL SIZES FOR CASING** When hand-nailing, the classic approach is to drive 3- or 4-penny nails into the jamb and 6-penny nails into the stud. When using a nail gun, it is common to use 18-gauge 2-in. nails for both jamb and stud. When driving into the jamb, angle the gun a bit to ensure that the nail will not poke out the jamb.

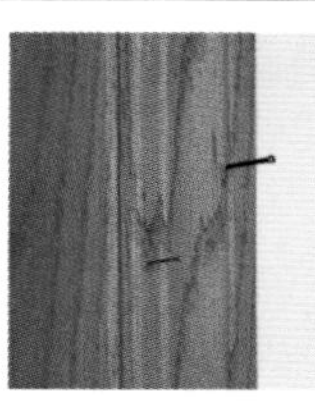

12 **NAIL FROM THE TOP DOWN.** Work your way down, driving nails and straightening the casing as needed as you go. Check the reveal before driving each pair of nails.

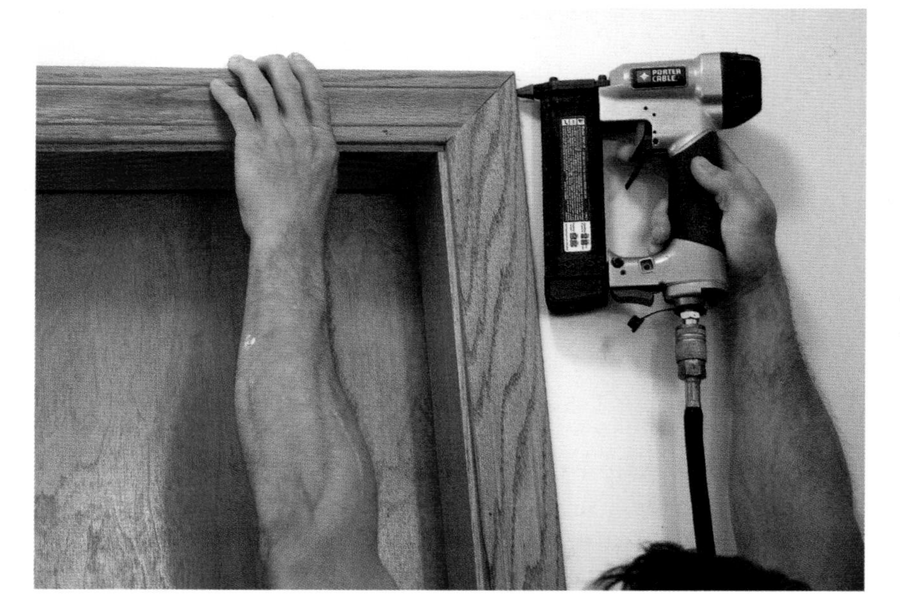

13 **REINFORCE THE CORNERS.** Reinforce each joint by driving two small nails through the edges of the top and side piece. You can use 23-gauge pin nails (or hand-drive very small brads), or, if you want more holding power, use 18-gauge nails with heads. Angle the nails slightly, so they will be sure not to poke out through the face of the molding.

TIP Take note of the positions of the studs on each side of the jamb. Usually they are within ½ in. or so of the jambs, but sometimes there is a significant gap between jamb and stud. Plan to drive nails that will hit the studs, not the gap.

Adding Backband

A quick and easy way to enrich the look of mitered casing is to add backband molding. The backband shown here has a simple chamfered inside edge; other types have a slightly more complicated profile.

1 **MARK THE MITER CUT ON THE BACKBAND.** Hold each piece in place and mark for cutting along the same miter line as the casing.

2 **CUT THE BACKBAND.** Be sure to hold the molding tightly flat on top of the saw's base.

3 **ATTACH THE BACKBAND.** Slip the backband pieces over the casing and attach to the casing with small nails driven both into the face (as shown) and into the sides.

Rosette and Plinth Block Casing

This style is one of the easiest casings to install; all the cuts are 90 degrees. Yet it produces a lush Victorian appearance. Fluted casing, as shown on the facing page, is a common choice, but any type of casing that is about 1⁄4 in. narrower than the rosettes and blocks will look handsome.

Choose a style of plinth block that doesn't clash with the base molding; the relatively plain style shown here goes with almost any style. The plinth blocks should be an inch or less taller than the baseboard.

ROSETTE DESIGN. Choose rosettes that are about 1⁄4 in. wider than your casing. In addition to the most popular ones sold at home centers and lumber yards, you can go to online sources for artistic rosettes like the ones shown here.

TIP Here we show installing trim onto a pretty rough surface in an older home—the owners chose to keep the older painted door and jamb and replace the casing only. Wide primed casing is a good choice for this situation: Gaps and imperfections can be covered with caulk.

1 **TEST THE FIT.** Hold or partially nail a scrap piece of casing on the side and the top and place a rosette between them. Determine where the rosette will look best. Often (as in this case) its bottom should be level with the bottom of the top jamb piece.

2 **MARK THE SIDE CASING.** Install a plinth block at the bottom, if there isn't already one there. Hold a piece of casing in place and mark it for cutting.

3 **ATTACH THE SIDE CASING.** Nail the casing to the wall and to the jamb, maintaining a straight reveal line.

4 **NAIL THE ROSETTE.** Apply some construction adhesive to the back of a rosette and attach it with a single nail, so you can slightly adjust its position if needed. Install the other side casing and rosette as well.

5 **NAIL THE HEAD CASING.** Measure between the rosettes and cut the top casing piece. Fit it snugly between the rosettes and drive nails to fasten to the wall and jamb.

Window Apron

If a window has a stool, it needs an apron beneath the stool. An apron is typically made of a piece of casing with its thick edge on top, just under the stool.

1. To find the apron's length, measure from outside to outside of the side casing pieces.

2. Cut the apron to length, with 45-degree bevels at each end.

3. To finish each apron end, make a return: Cut a scrap piece of casing at 45-degree miters on each end, and cut off the bevels as shown.

4. Attach the returns with glue and pin nails.

5. Center the apron under the stool and nail in place.

1. FIND THE APRON'S LENGTH.

2. MAKE BEVEL CUTS.

3. CUT THE RETURNS.

4. ATTACH THE RETURNS.

5. ATTACH THE APRON.

Casing with a Fillet

This traditional-looking style features a thin protruding horizontal piece just under the top casing and on top of the side casing pieces. Variously called a "fillet," "parting bead," or even "popsicle stick" because of its thinness, this piece adds a distinctive touch for a small outlay of money and labor. A fillet may be a simple piece of lath; in the example shown here, we added some detail by using a piece of upside-down Colonial stop.

Fillets are usually used along with side casings that are not tapered along their widths, as most mitered casings are. The side casings could be plain, without a profile (sometimes called "sanitary casing"). In our example, we add a decorative touch to the side casings by attaching outside corner molding to each side; we left the head casing piece plain.

FILLETED CASING. This trim arrangement is made with plain (or sanitary) side casing that is trimmed on the sides with outside corner moldings, a ripped 1×6 for the head casing, Colonial stop, and plinth blocks.

TIP It's important that both the side casing pieces be precisely tall enough to create the desired reveal on the top jamb. Before fastening the side pieces, set a scrap piece of wood on top, spanning both pieces, to be sure the reveal will be consistent.

1 ASSEMBLE THE SIDE CASING. To make this style of casing, nail pieces of ¾-in. outside corner trim to each side of 3½-in. sanitary casing. Hold a straightedge at one end as shown.

Making Plinth Blocks

If you cannot buy plinth blocks that suit you, it's not difficult to make your own. In our example, none of the available blocks came close to matching plinth blocks elsewhere in the house.

You may have to go to a hardwood supplier to buy some 1-in.-thick or $1\frac{1}{8}$-in.-thick stock.

Find a router bit (**1**) that creates a similar profile; you probably won't get an exact match, but the difference will not be noticeable if the blocks are in different rooms. Cut the blocks to length and width (**2**). Use a power sander to smooth any rough edges, then rout the top and side front edges (**3**).

1. FIND A ROUTER BIT WITH A SIMILAR PROFILE.

2. CUT THE PLINTH BLOCK TO LENGTH.

3. ROUT THE PROFILE.

TIP If the baseboard is in place, hold a plinth block (this one is scheduled for edge routing) in place to mark for cutting the baseboard. Note that the block is wider than the casing, and so nearly covers the entire thickness of the jamb.

2 **MEASURE FOR AND CUT THE SIDE CASING.** Install the plinth block at the bottom, if it will be used. (Of course, for window trim the casing will rest on the stool.) Mark the reveal, as shown on p. 124. Measure for side casing pieces that reach the top reveal mark, and cut the casing.

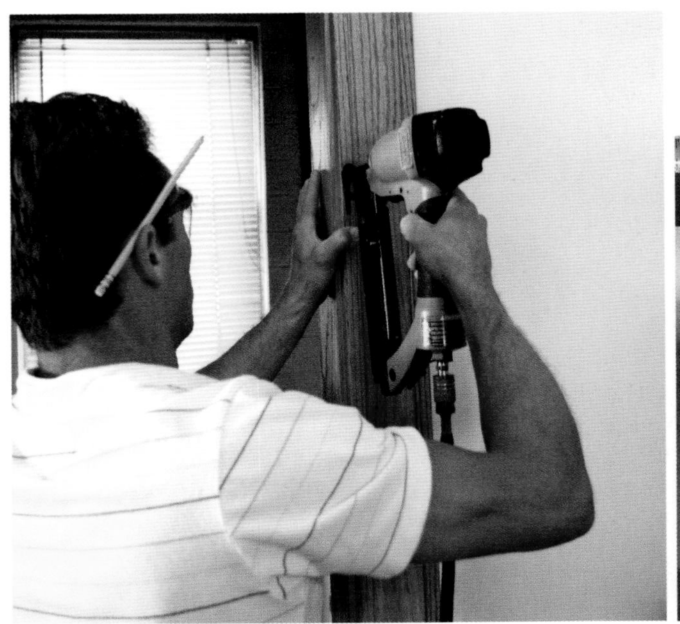

3 **ATTACH THE SIDE CASING.** Be sure to maintain a consistent reveal along the side and top jamb edges. Drive pairs of nails, one into the jamb and one into the wall framing, every 16 in. or so.

4 **MEASURE FOR THE FILLET.** Measure across the outside edges of the casing boards and add 1 in., so it will overhang ½ in. on each side.

5 CUT RETURNS FOR THE FILLET. Cut a piece of 1⅜-in. Colonial stop to length with 45-degree miters at each end. (Measure from tip to tip of the miter cuts.) Place the stop upside down on the saw's base to make the cuts. To make returns, cut 45-degree miters on both sides of a scrap piece of stop, then cut across at 90 degrees as shown to make a triangle.

Modifying a Fillet for an Imperfect Wall

If your wall is wavy or otherwise imperfect (a not unusual situation, especially in an older home), you may need to shave one end of a fillet so it can come tight against the top jamb. Use a plane for a small discrepancy, or make a cutout with a tablesaw and handsaw (below) to close a gap of more than ⅛ in.

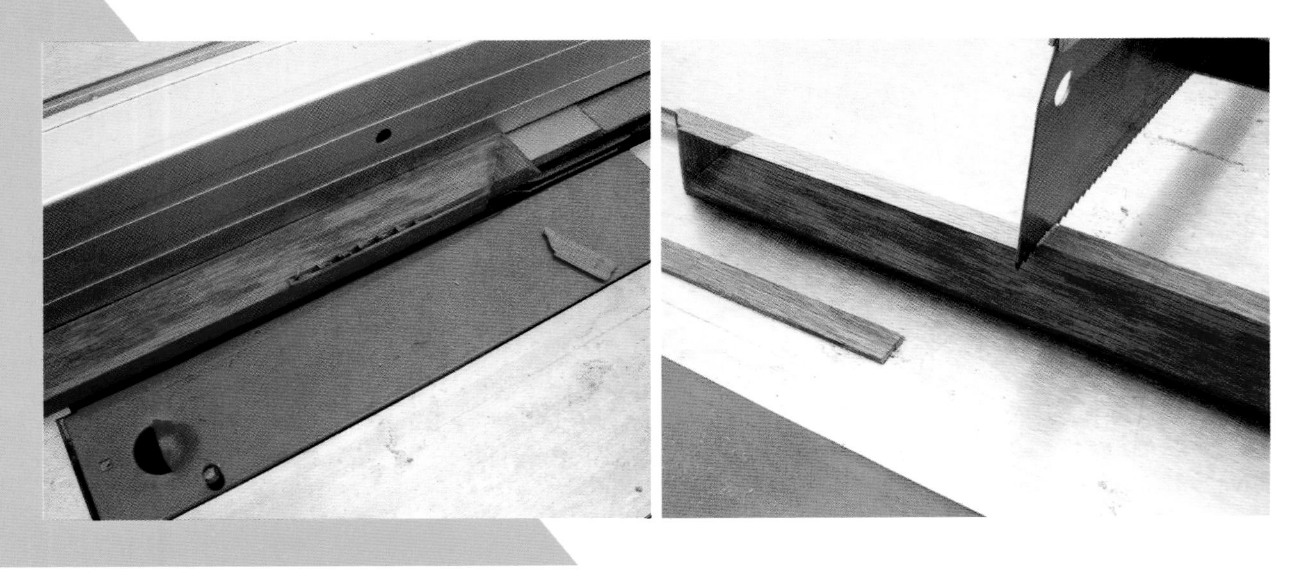

6 **GLUE AND NAIL THE RETURN.** Test that the triangular return piece fits neatly. Apply glue to the return piece, press it into place, and drive a pin nail or two to hold it until the glue dries.

7 **NAIL THE FILLET.** Place the fillet atop the side casing pieces. Check that it comes tight to the edge of the top jamb; if there is a gap, see the sidebar on the facing page to solve the problem. Position the fillet so it overhangs the casing on each side by ½ in., and drive small nails down into the top of the casing.

8 **ATTACH THE HEAD CASING.** Rip-cut a piece of 1×6 to the same width as the side casings (including the attached outside corners)—about 4¼ in. Cut it to the same length as the distance from outside to outside of the side casings. Set it centered on top of the fillet. Drive nails up through the fillet and into the head casing; nail the head casing to the wall framing as well.

Picture-Frame Window Casing

This type of molding has no stool or apron—just four casing pieces that frame the window as if it were a work of art. Picture-frame casing is often used for casement windows, where a stool would not be very usable anyway, but it's not uncommon for sash windows in a modern house.

In theory, this type of casing is simplicity itself: Just cut four pieces to the correct lengths, all with 45-degree miters, and nail them up. And if your window jamb is perfectly square and your saw cuts perfect miters, installation might go just that smoothly. However, things are rarely perfect in this world, and if an angle is off only slightly, with picture-frame molding the imperfections are multiplied by four. So with the following steps, proceed carefully to ensure you keep all the joints nice and tight.

1 **CHECK FOR SQUARE. As with any casing installation, check the window jamb for square. For picture frame, inspect all four corners especially carefully, noting even tiny imperfections. As much as possible, adjust shims to make the corners square; time spent doing this will save installation time and help produce a neater job.**

2 **MARK THE REVEAL. Use a plywood gauge guide (p. 124) or a sliding square to mark the desired reveal on all sides.**

3 **MEASURE FOR ALL FOUR PIECES. If the measurements for the two sides or the top and bottom differ, adjust the jamb to make them equal.**

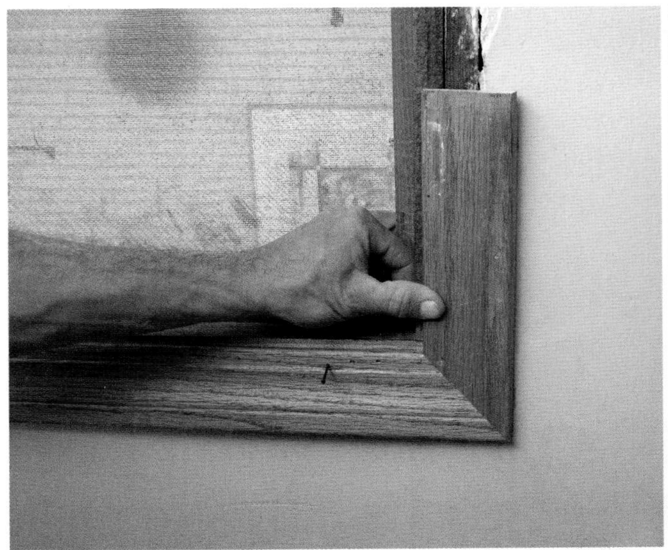

4 **TEST THE ANGLE.** Cut two scrap pieces of casing at 45 degrees, and hold them at all four corners to be sure they follow the angle of the jamb. If not, you may need to adjust your miter saw, as shown on p. 56.

5 **FIRST PIECE.** Cut one end of a piece for the top or bottom. Measure for cutting the other end by burning an inch, as shown. Test the fit on both the top and the bottom.

6 **SECOND PIECE.** If you are satisfied, use the first piece as a template to mark the tip (only the tip) of the second piece.

7 **TACK TOP AND BOTTOM.** Use a hammer to "tack" the top and bottom pieces in places so they are aligned with the reveal lines. Partially drive two or more nails so you can remove them if you need to adjust the casing.

8 **MEASURE FOR THE SIDE PIECES.** This measurement should be the same on each side.

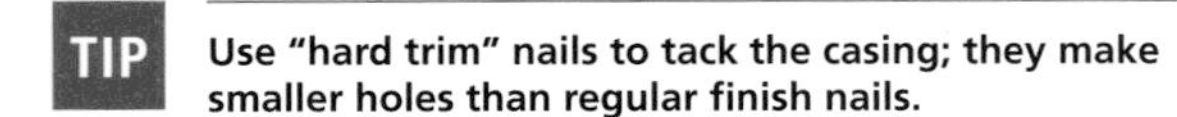

TIP Use "hard trim" nails to tack the casing; they make smaller holes than regular finish nails.

9 **ADJUSTING THE FIT.** Cut a side piece $1/8$ in. or $1/16$ in. longer than your measurement. Hold the board in place and test the fit. You may find that it only needs to have one end cut slightly to length. Or you may see that an angle or two needs to be changed very slightly. Make the adjusting cut with the miter saw. See pp. 58–59 for tips on micro-adjusting angles.

TIP If you have a handsaw rather than a power miter saw, making micro-adjustments will be nearly impossible. In that case, it's best to try to make the first cut accurate. Then make small adjustments with a plane or sander, as shown on p. 73.

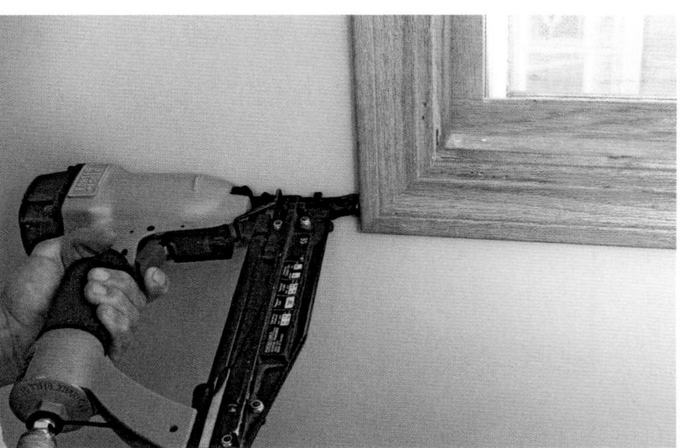

10 **INSTALL THE SIDE PIECES.** You may need to remove one or more nails to adjust the position of the top or bottom piece in order to make a tight joint. Once you are happy with the joints, drive finish nails to attach the casing all around. Drive a pair of nails, one into the jamb and one into the wall framing, every 16 in. or so. If you are hand nailing, use 3d hard-trim finish nails for attaching to the jamb and 6d nails for attaching to the wall.

Casing in a Tight Spot

Not all doors and windows have ample wall space around them, allowing you to install full-width casing. Some are in tight spaces, so that the head casings and sometimes even the side pieces bump into other casings.

Depending on the type of trim and how close things are, you may choose to emphasize the individuality of, say, two doors, and narrow the trim of each. Or you may choose to treat the two elements as a single unit, installing what can look like a single shared top or side casing that looks like one piece.

With either approach, it's important to keep casing heights precisely the same; if two nearby casings have heights that differ by even ⅛ in., the trim will look awkward and sloppy. If the doors or windows are themselves misaligned in height, it may be worth your while to move or even change one of them. If that is not feasible, give priority to the overall appearance of the trim—for instance, make the top casings end up at the same height, even if that means one casing will be wider than the other.

Walls are rarely perfectly square or plumb, so this type of work often involves scribing lines and making rip cuts that slightly slope in order to make pieces fit together tightly on misaligned walls. This can be painstaking work.

The photographs here were taken in a butler's pantry, which has several nearby doors facing in different directions, as well as a base cabinet to negotiate around.

Abutting side casing pieces

At this very tight corner, two side casings will be installed tight together at right angles. Only one side of each of the pieces will receive a finish treatment (which in this case is a piece of outside corner molding), creating a sort of illusion of a single wide casing piece.

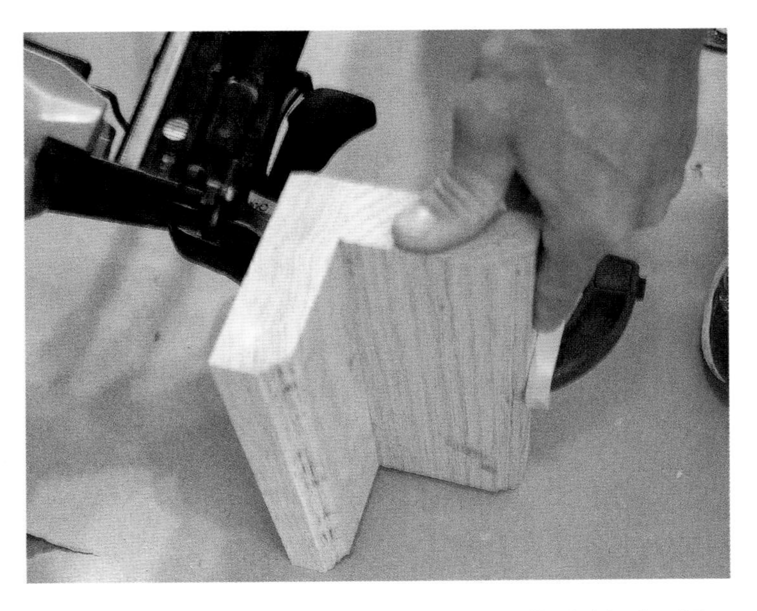

DOUBLE PLINTH BLOCK. At a corner, make a plinth block with two pieces fastened together. Rout the edges of only one side of each piece. (This doubled plinth is seen in the top left photo on p. 142).

1 MEASURE FOR THE WIDTH OF THE CASING. **Measure in several places for cutting the first side piece to width.**

2 **PLINTHS AND ONE CASING INSTALLED.** The first piece does not need to fit tightly, so you can make a straight rip cut on a tablesaw at the narrowest measurement. Install the first piece, in this case atop a double plinth block.

3 **SCRIBE FOR THE SECOND PIECE.** The second piece must be cut precisely so one side follows the jamb's reveal line and the other fits tightly against the first casing piece. Cut the casing to length. Find two scraps of wood of the same thickness as the desired reveal line on the jamb. In this case, the scraps are as thick as the reveal line plus the thickness of the outside corner molding that will be installed. Tack the scrap pieces in place, and press the casing against it, upside down as shown. Scribe a cut line along the jamb.

TIP It's easy to get confused when installing custom casing like this. After you have scribed the line for the second piece, double-check the measurement with a tape measure.

4 **RIP THE CASING.** If the scribed line is not parallel to the side of the board (and it likely will not be), you'll need to rip-cut a line that angles slightly, and you cannot use a tablesaw's fence. Instead, cut with a circular saw, as shown. You'll need to clamp the board at the front, then at the back, as you progress along the cut. If the cut is not straight, use a plane, a power plane, or a belt sander to fine-tune it.

5 **ATTACH THE SECOND PIECE.** Press the casing board in place. If you see any gaps, use a plane to remove any wide spots and so achieve a tight fit all along the joint. Nail the side casing in place.

Head casing

Check the side casings with a level to be sure they are at exactly the same height, so the head casings will be perfectly aligned. You may need to cut one or more top casing piece longer or shorter than usual in order to achieve a seamless appearance.

1 **ALIGN THE FILLETS.** At an inside corner, cut fillets a bit longer than they need to be, at 45-degree miters where they will meet. Butt the pieces together and check the joint; you may need to adjust the miter to make a tight fit. Once they fit tightly at the corner, cut them to length.

2 **INSTALL THE TOP CASING.** Head casing pieces can usually be butted together without miter or bevel cuts. Cut them a bit long, butt together, and check the fit. If needed, scribe a line on one of the pieces and cut to produce a cut that fits tightly. Cut the pieces to the correct length, and fasten.

MORE CASING RECIPES

Here are some additional popular casing configurations, ranging from the simple to the complex. All can be made with available moldings: Most can be found at home centers and lumberyards, though you may need a specialty hardwood store or an online source for a few of the others.

SIMPLE BUTTED CASING

This straightforward casing harmonizes well with a cottage or Arts and Crafts décor. On the sides, run plain ("sanitary") casing, which is ½ in. thick and 3½ in. wide, up to the reveal mark on the top jamb. Cut a head casing of 1×4 (¾ in. thick and 3½ in. wide) so it overhangs the side pieces by ¾ in. on each side, and attach.

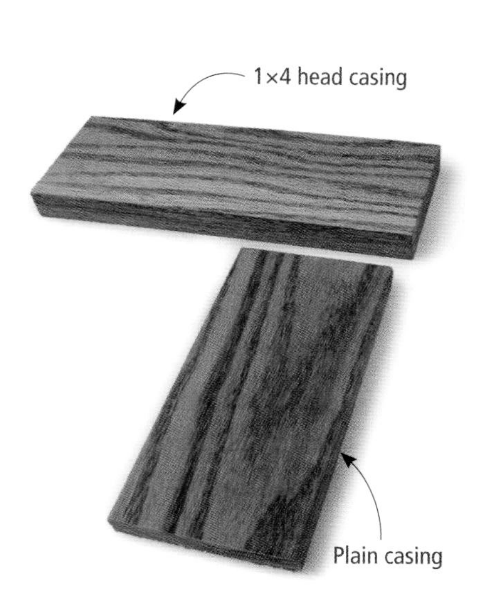

ARTS AND CRAFTS

The term "Arts and Crafts" means different things to different people, but most would agree that this arrangement fits the category. It's much like the Simple Butted Casing (above), but the side pieces are made with 1×4 and the head casing is made with 5/4 stock (which is 1 in. or 1⅛ in. thick). Cut the head casing's ends at a 5-degree angle (or a 95-degree angle, depending on how you look at it).

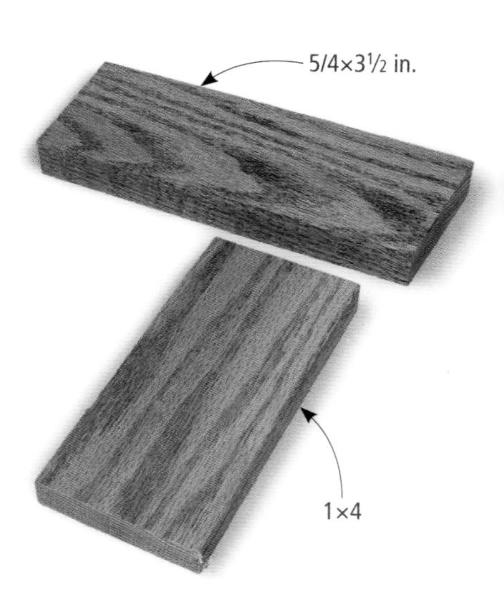

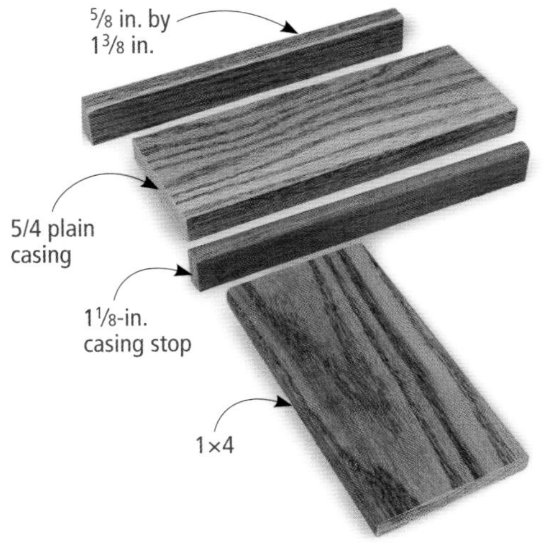

TRADITIONAL

This design is also easy to install, with no mitered joints or returns, but a fillet and top cap give it extra texture. Attach plain casing to the sides. Add a fillet made of 1 1/8-in. casing stop, which has a rounded front edge and is 1/2 in. thick. If you cannot find casing stop, rip the piece to fit. Add a head casing of 1×4, and nail the fillet up to it. Top it all off with a piece that is 5/8 in. thick and 1 3/8 in. wide; you may need to cut this yourself with a tablesaw. The fillet and cap in the example shown have just slightly rounded front edges, and the ends are cut straight. You may choose to round the ends as well.

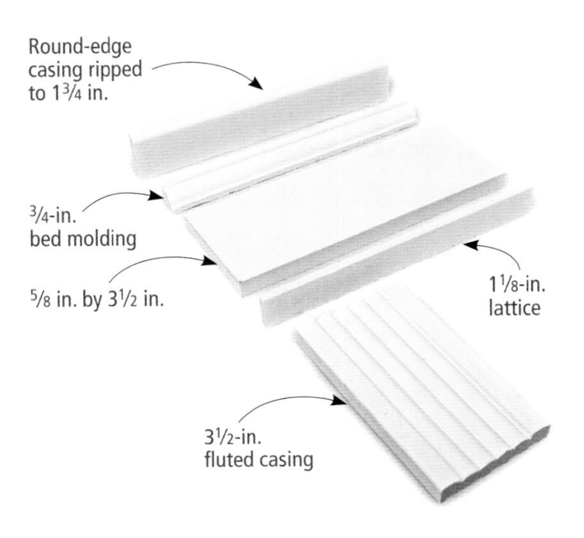

CROWNED TOP

Putting a small crown detail on top of casing is a typical Victorian touch. It adds plenty of old-fashioned class and is actually not all that difficult to achieve. Install plain side casing, topped with a fillet—here, a piece of casing stop, though you could use another thin piece such as upside-down Colonial stop. Set a 1×4 on the fillet, and trim its top with either a small crown molding or, as shown here, bed molding, which looks like small crown but is a solid piece, so there will be no openings on top of the casing. You will need to cut return pieces for the sides (see p. 71). Top it all with a thin piece; here, we use upside-down ranch casing that has been ripped to fit.

CLASSICAL

Whimsical yet stately, this classical (or neo-classical) look features fluted side casings reminiscent of Greek columns and a head casing crowned with straightforward (as opposed to Victorian) elements. Molding like this is usually painted, and is often made with MDF pieces. Install the fluted side casings, and attach lattice on top so it overhangs 1/2 in. or so. Rip-cut the 5/8-in. by 3 1/2-in. "frieze board" out of a wide base or other molding. Wrap its top with a simple bed molding; cut a return for the sides. Top the whole thing with round-edge casing ripped to 1 3/4 in. or so. Paint carefully, for a smooth surface.

CHAPTER EIGHT

CROWN MOLDING

CROWN MOLDING is so called because it sits at the top of walls, snug against the ceiling. It may also be thought of as a crowning touch for a room, because it is highly visible and adds plenty of style points in one broad stroke. Traditional crown molding has multiple curves for a richly textured, Victorian look, but sleek modern crown options are also available.

Crown molding has two narrow flat sections at the top and bottom of its back, which rest against the wall and the ceiling. Most crowns are designed to be taller than they are wide once installed.

A single piece of crown molding can make a big impact, especially if it is wide. For even more visual pop, consider a built-up crown (also called a cornice) composed of two or three pieces; see pp. 166–169.

Crown is probably the most difficult type of molding to install, so we devote a chapter to helping you get all the joints tight. Don't rush. Take time to test the corners with scrap pieces, and fine-tune your cuts where needed. Measure and mark walls to be sure you will easily find the studs and joists as you attach it, and install blocking where needed.

Installing hardwood crown calls for extra care and precision. Primed or PVC crown is a bit easier to install, because minor gaps can be filled with caulk.

Crown Options

The width of the crown should be in proportion to the room's height. If you have an 8-ft.-tall ceiling, it's generally best to avoid moldings that drop more than 5 in. down from the ceiling. For a 9-ft. or 10-ft. ceiling, the drop can be up to 7 in. before it starts to seem too dominant.

A Picture molding is placed about 4 in. below the crown, and the space between is painted the same color as the moldings above and below. This produces the illusion of very wide crown molding. B Simple, narrow crown molding is no less elegant for being understated. C The top-of-wall molding here is not actually a crown, because it doesn't curve out onto the ceiling. Instead, wide flat molding pieces simply reach up to the ceiling, for a slightly more modern look. D Wide crown molding in combination with a band of chair rail–like molding that runs about 3 in. below creates a handsome, classic effect. E This built-up molding has modest detailing at the bottom and at the ceiling and features a wide, smooth cove in the middle—in keeping with the other moldings in the room, which also are more flat than detailed. F In keeping with the Craftsman style of the casings below, this crown has a simple large cove reaching up

D

to the ceiling and a detail below that makes a small shelf. Crucial to the look is the natural oak wood, stained a medium-dark color. **G** Two pieces of molding—a wide crown at the top and a chair rail at the bottom—are perfectly joined at the inside corner, neatly stained a chocolate brown, and given a glossy finish; the simple, clean lines would work with a modern or retro theme.

E

F

G

SIDE VIEW OF CROWN MOLDING

Crown molding is installed at a "spring angle," so it drops from the ceiling farther than it extends out from the wall. The crown's most detailed side is positioned at the bottom for best visual appeal.

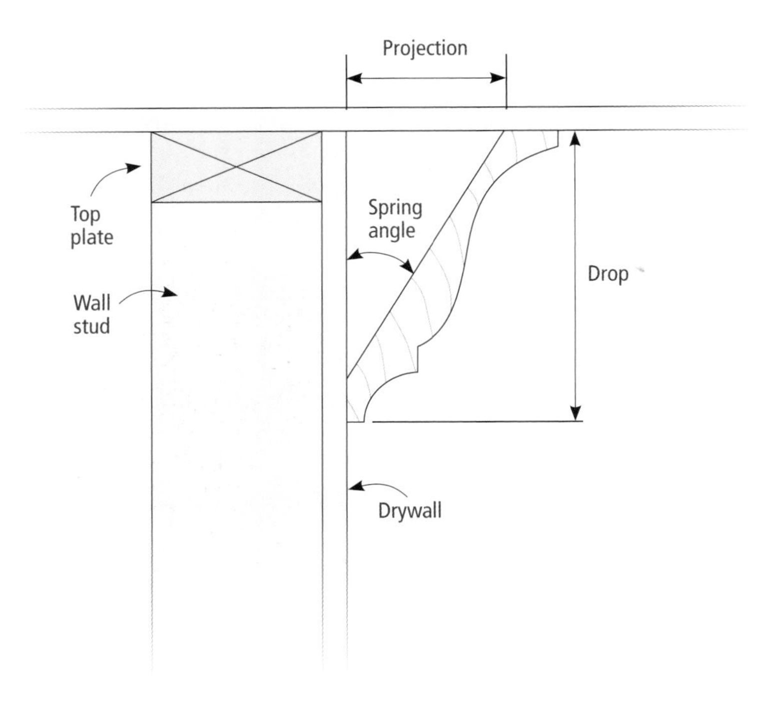

Prepping the Room for Crown

More than with any other type of molding, it pays to spend extra time planning a crown-molding job, choosing materials, and preparing walls, ceilings, and even your miter saw. Calibrate your power miter saw (p. 56) to be sure it cuts with maximum precision. For long pieces, you'll get better results if you have a helper hold one end while you check the fit or fasten into place.

Wall and ceiling conditions

Because it is installed at an angle, crown cannot be bent to follow wall or ceiling contours as readily as base molding. If your walls or ceiling have noticeable waves or curves, the crown will probably need to be caulked in places, perhaps in ways that are noticeable. If you are able to fix waves or out-of-plumb walls with joint compound, take the time to do so.

Using Inexpensive Hand Tools

Crown can be cut with a hand miter saw, as long as it's not too wide. Cutting will be somewhat difficult and slow. Use clamps to hold the board firmly in place as you cut. Installing with hand-driven nails is a bit difficult, largely because it usually takes two hands. To solve that problem, drill pilot holes first, insert the nails into the holes; then you can hold the board in place with one hand and hammer with the other.

Finding studs and joists

Unless the crown is very small, nails will not be able to reach the wall frame's top plate, so you need to find and attach to wall studs and ceiling joists. Ceiling joists run in one direction. You probably will be able to attach to ceiling joists on only half the walls; on the other half, the joists run parallel to the wall and usually cannot be reached with nails. On those walls, you may need to install blocking for nailing surfaces (see pp. 152–154).

TIP **Especially if a ceiling is wavy, some installers find it helpful to chalk a line on the wall indicating the bottom of the installed crown. To do this, scribe lines at each end, then chalk a line between them. Using a scrap piece of crown, check that the molding will not go below the line at any point; if it will, chalk a lower line.**

TIP **Use a felt-tipped pen and press lightly on the tape; if you press hard with a pencil, paint may get pulled away in those places when you remove the tape.**

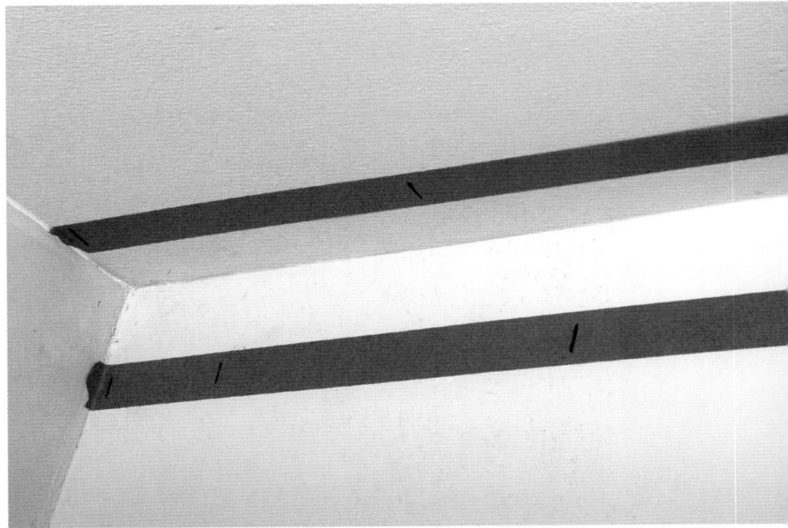

SQUARE UP CORNERS. Inside corners of walls are often slightly rounded, which can make it difficult to install crown molding tightly. Use a chisel to square up an inside corner.

TAPE AND MARK STUDS. Use a stud finder to mark locations of studs and ceiling joists in a way that will be visible when the crown molding is held in position. If you will paint the wall or ceiling anyway, go ahead and mark with a pencil. If you want to keep the surfaces clean, apply painter's tape just beyond where the crown will be and mark the tape.

Planning the install

If possible, plan an installation so you don't have to make a cope cut on both ends of a piece; it's not impossible to make such a piece, but it's a challenge to get it just the right size and with tight joints on each end.

Blocking

Crown molding should be nailed firmly into wood framing at the top and bottom. Avoid the temptation to just nail to drywall; doing that may look OK at first, but it can come loose in time. For that reason, it's always a good idea to install blocking where the ceiling joists run parallel to the wall, and thus where there are no ceiling joists to nail to. Many carpenters install blocking on all the walls, to be sure they will have a solid nailing surface wherever they need it.

Blocking is for nailing purposes only. Because blocking follows the imperfect lines of walls and ceilings, blocking cannot be a reliable guide for setting the crown molding at the correct angle. In fact, blocking should be cut a bit small, so there is a slight gap between it and the back of the crown. You may install individual blocking pieces, or long continuous blocking.

Individual blocking triangles

Because of the crown's spring angle, you cannot simply cut a series of blocks at a 45-degree angle. You can follow the chart and instructions on p. 159 to find your crown's spring angle, or use the method shown on the facing page.

TIP **If a wall is too long for a single crown molding board, you'll need to make a scarf joint (p. 161). That can actually simplify the installation: Each of the two pieces can have a miter or cope cut at one end. Cut the second of the pieces a bit long, check that the miter or cope cut is nice and tight, and then mark it for cutting to length at the scarf joint.**

ORDER OF INSTALLATION FOR CROWN MOLDING

In this example, the first piece has an outside miter cut at one end and a butt cut at the other end. Cuts 2 through 5 all have one butt cut and one cope cut. The last piece, number 6, is the most difficult, with a miter cut on one end and a cope cut on the other. You can make the cope cut first and check it for a tight fit, then hold in place to mark for cutting the outside miter on the other end.

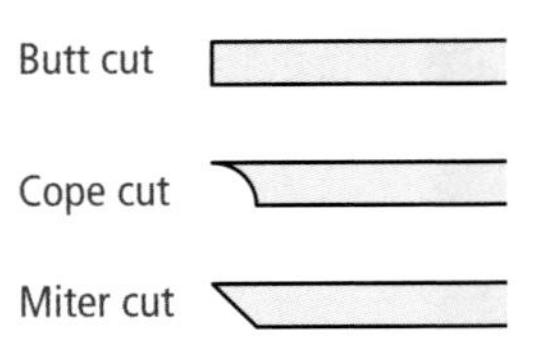

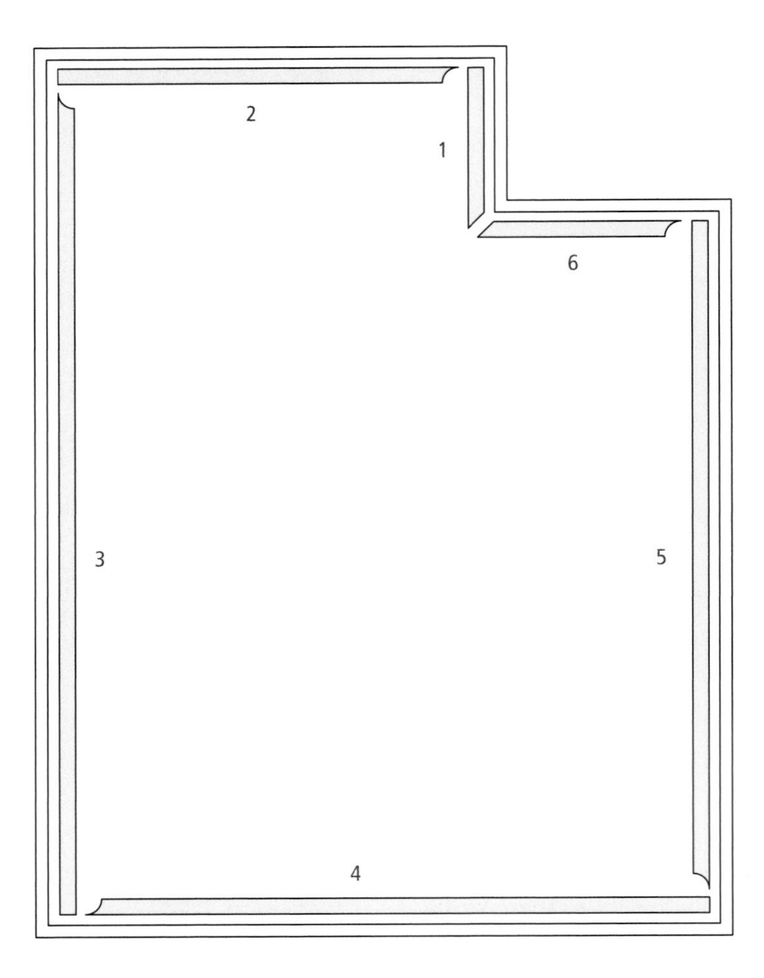

1 **ESTABLISH THE BLOCKING MEASUREMENTS.** With a piece of crown molding set upright on a table, position a framing square to represent the wall and ceiling, with each of its legs against the short flat sections at the top and bottom. This gives you the protrusion and drop measurements.

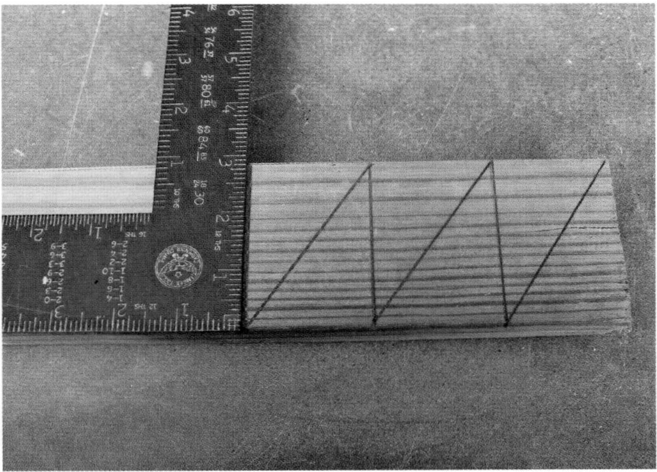

2 **MARK BLOCKING TRIANGLES.** Rip-cut a 2×4 to the "drop" measurement. Along one edge of the ripped board make marks for the "protrusion" dimension, and use a straightedge to draw square and connecting lines.

3 **TEST THE FIT.** Cut through the lines with a chopsaw. The thickness (or kerf) of the sawblade will make the pieces slightly smaller than the dimensions—which is what you want. A finished block should be slightly smaller than the back of the molding, as shown.

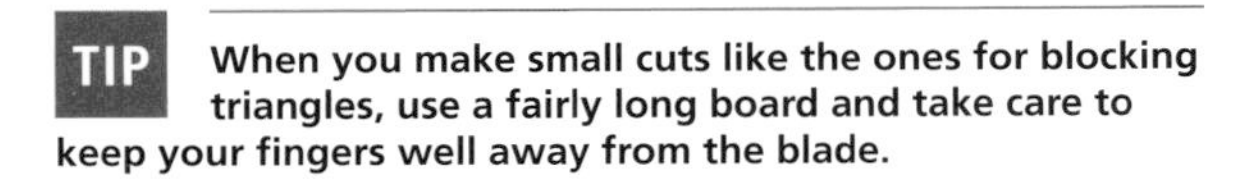

TIP When you make small cuts like the ones for blocking triangles, use a fairly long board and take care to keep your fingers well away from the blade.

4 **ATTACH THE BLOCKS.** Drill pilot holes and drive small-headed screws to attach the blocks to wall studs or to the top plate.

Continuous blocking

Installing a single piece of continuous blocking ensures that you can fasten the crown molding wherever you want.

1 **DRAW THE BLOCKING PROFILE.** On a piece of paper, draw two perpendicular lines to represent the wall and ceiling. Set a piece of the crown against the two lines, as it will be installed. Scribe a line on the back of the molding.

2 **DRAW CUT LINES FOR THE BLOCKING.** Set a piece of 2× lumber (in this case, a 2×3) on edge, with its back and top aligned with the wall and ceiling lines. Draw short cutlines where the angled line intersects with the board.

3 **MAKE THE ANGLED RIP CUT.** On the front edge of the board, draw a line connecting the two lines from Step 2. On a tablesaw, raise the blade and set its angle to match the line; it should cut to the inside of the line, so the block will be about ⅛ in. smaller than the opening behind the crown. Rip-cut the board.

4 **BLOCKING IN PLACE.** There should be a small gap between the block and the back of the crown molding when correctly installed. Nail or screw the block to wall studs or to the framing's top plate.

Setting Up a Chopsaw

You can cut crown molding simply by positioning it (probably upside-down) at its spring angle on the chopsaw's table and fence. However, the two flat sections at the back of the molding, which rest against the wall and the ceiling, are often only ½ in. or ¼ in. wide, so it's easy for the molding to slip out of alignment. (It can even be difficult to get the molding in correct alignment to begin with.) For that reason, carpenters often set up their chopsaws with jigs, guides, or at least layout lines to ensure that the board is being held correctly during the cut.

Also, your fence may not be tall enough—on one or both sides—for your crown molding. In that case, you may choose to cut the molding lying flat if your chopsaw has a bevel feature (see p. 59). Or attach a piece of plywood to the fence, or an L-shaped jig, as shown on p. 156.

Marking lines

A simple line may be all you need. You can make the layout line on the chopsaw's table or its fence, or both.

TIP **The type of chopsaw setup you choose depends on a variety of factors: the width of the molding, the height of your fence (on both sides of the blade), and your confidence in holding the molding in correct alignment.**

GUIDE LINE ON TABLE. To make a guide line on the saw's table, apply pieces of painter's tape on the table. Press a piece of the crown molding in position (upside down, with the detailed edge up) and check to be sure its flat edges are flat against the fence and the base. Scribe a line on the tape.

GUIDE LINE ON FENCE. Attach a strip of plywood to the fence. Hold the crown in position, and scribe a line on the plywood.

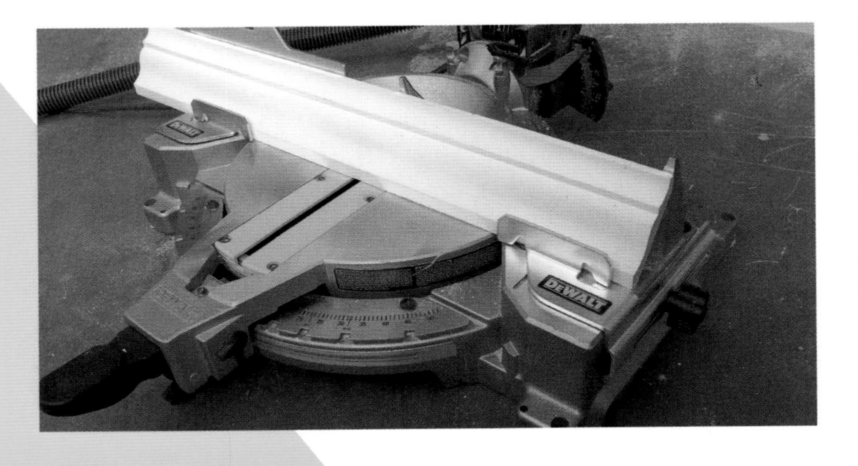

Using a Crown Stop

Some miter saws are provided with crown stops, whereas others will accept optional crown stops that can be attached to the base of the saw. Position the crown molding correctly, slide the stop against the molding's bottom, and tighten the mounting screw.

L-shaped crown jig

An L-shaped plywood jig not only helps hold the molding in position but also makes it easier to cut to the correct length when the cutline is at the bottom. Making and installing this fence calls for a bit of trial and error but will not take much time.

TIP You will need to make 90-degree cross cuts as well as miter cuts. With the L-shaped jig, be careful not to cut fully across, or you will cut the jig in two.

1 **ATTACH THE PLYWOOD.** Rip-cut two pieces of plywood to about the widths of the fence and the table. Here we use ½-in. plywood, which allows for a bit more cutting room, but you can use ¾-in. plywood. Drive screws through the back of the saw's fence to attach it.

2 **CUT MITER LINES.** Make miter cuts in the jig as far as you can. In this case, the left-hand miter can be cut to full width, but on the right side the saw's motor bumps onto the fence and limits the cut.

3 **MAKE CUTOUTS IN THE JIG BACK.** Mark where cutouts are needed so that the saw can descend far enough to make full miter cuts on each side. Remove the jig and cut using a jigsaw. Reattach the jig and test; you may need to make further cuts.

4 **ATTACH A STOP STRIP.** Position the crown molding against the jig as you will cut it (with the bottom side up). Check the backside of the molding to be sure its flat surfaces are pressed against the fence and table. Press a thin strip of wood to act as a stop against the bottom of the molding, and attach with pin nails or staples.

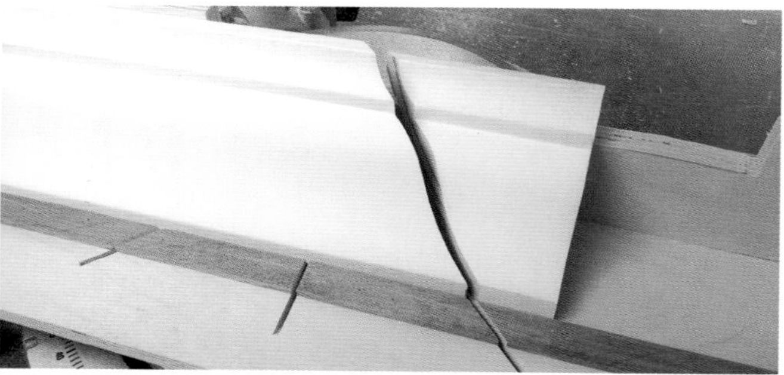

5 **CUTTING WITH THE STOP STRIP.** Cut through the strip at both miters; at the 90-degree crosscut cut only partway through, so you don't cut the jig in half. Now, when cutting a miter for an inside corner, you can easily align the board against the cutline.

Understanding the Cuts

Until you are practiced at it, cutting crown molding for inside and outside corners can be confusing. And making a wrong cut can mean having to throw out an expensive board. But once you've gotten a few simple ideas seared into your brain and you've developed some correct habits, cutting crown will be easier.

The right orientation

In most cases, crown molding should be installed with the detailed edge down (see the drawing on p. 150). Installing it that way will make the crown taller and more visible. In unusual circumstances, such as our mantel project, the crown may be installed upside down, with the detail up (see the photo on p. 214).

Also, it's important to position the crown at the right angle. Unfortunately, crown easily slips against the wall and ceiling, which takes it out of alignment and throws joints out of whack. To help avoid this slippage, you may choose to chalk a line on the wall showing the lower edge of the crown. Another way to keep things correctly angled is, whenever possible, to hold both miter-cut pieces together—or hold one cut piece and a scrap piece for the other side—and make sure you've got a nice tight joint before driving any nails.

TIP **Just to be crystal clear: When talking about the four crown cuts, the directional refers to what you see when looking at the joint. For instance, a piece that is "left side of an inside miter" appears on the left side of an inside corner as you view it.**

Make Samples for Reference

There are four basic miter cuts, for the left and right sides of inside and outside corners. Whichever method you use to cut, it helps to have clearly marked samples of all four on hand. Before making a cut, lower the miter saw's blade and hold a sample against it to be sure the saw is at the correct angle and the board is in the correct position.

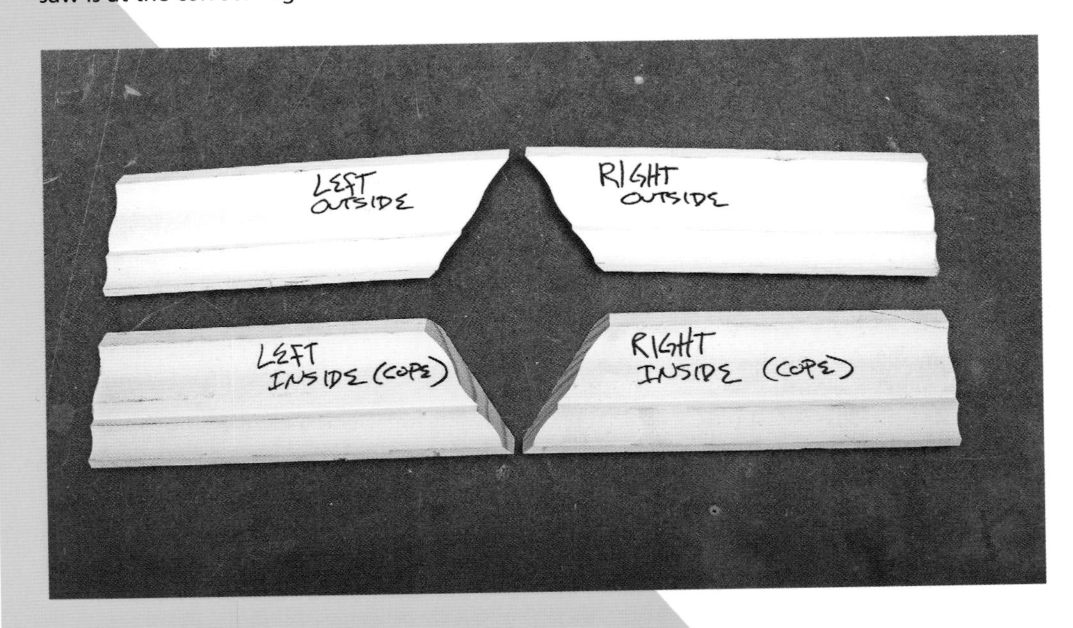

Cutting with simple miter cuts

Just about any crown molding can be cut with a simple miter saw, without a bevel feature. See pp. 155–156 for equipping your saw with a jig, guide, or markings to keep the molding at the correct angle while cutting.

Position the molding upside down, with the top side on the table and the bottom (more detailed) side against the fence. This may seem confusing at first, but will take only a few cuts to get used to.

RIGHT OUTSIDE. **To cut the *right* side of an *outside* miter, adjust the blade to a 45-degree angle to the right and place the board to the left.**

LEFT OUTSIDE. **To cut the *left* side of an *outside* miter, adjust the blade to a 45-degree angle to the left and place the board to the right.**

LEFT INSIDE. **To cut the *left* side of an *inside* corner (in preparation for making a cope cut), adjust the blade to a 45-degree angle to the right and place the board to the right.**

RIGHT INSIDE. **To cut the *right* side of an *inside* corner (in preparation for making a cope cut), adjust the blade to a 45-degree angle to the left and place the board to the left.**

Cutting with a compound-miter saw

If your miter saw—whether sliding or not—can be adjusted for bevel as well as angle, you can cut crown molding as it lies flat on the saw's table. If your crown is very wide, this may be the only way you can cut it. When cutting this way, sometimes the board's bottom edge is against the fence, and sometimes the top edge is against the fence.

Different moldings have different spring angles. First find your molding's spring angle, and then use the chart at left below to set the saw's angle and bevel.

> **TIP** **Because many crown moldings have a spring angle of 38 degrees, some miter saws have a cutting angle stop notch at 31.62 degrees—which is the ultra-precise cutting angle for that type of crown.**

1 **FIND THE SPRING ANGLE.** **The molding's spring angle is the angle at which it comes out from the wall—not from the ceiling. Use a protractor-type tool like the one shown to find the angle. Hold it at the bottom of the molding, where the smallest detail is. Most crown molding has a spring angle of 30, 35, or 38 degrees, but other angles are possible.**

ANGLE SETTINGS FOR CUTTING CROWN LAID FLAT

CROWN'S SPRING ANGLE	MITER	BEVEL
30°	27°	38°
35°	30.5°	35°
38°	31.5°	34°
40°	33°	33°
45°	35°	30°
52°	38°	26°

2 **SET THE BEVEL AND ANGLE.** **Consult the chart at left below and set the saw's bevel and angle. In this example, the spring angle is 35 degrees, so the angle is set to 30.5 and the bevel is set to 35. The bevel setting will remain the same for all cuts, but you'll need to change the angle adjustment often as you cut the different pieces.**

3 **CUT RIGHT OUTSIDE.** To cut the *right* side of an *outside* miter, adjust the blade angle to cut to the right. Position the board on the right of the blade, with the *top* edge against the fence.

Butt Joints

Many older homes have crown molding with visible butt joints. These have a hand-made charm and are less demanding to execute: Cut both pieces at 90 degrees, butt them together, and it will look good enough.

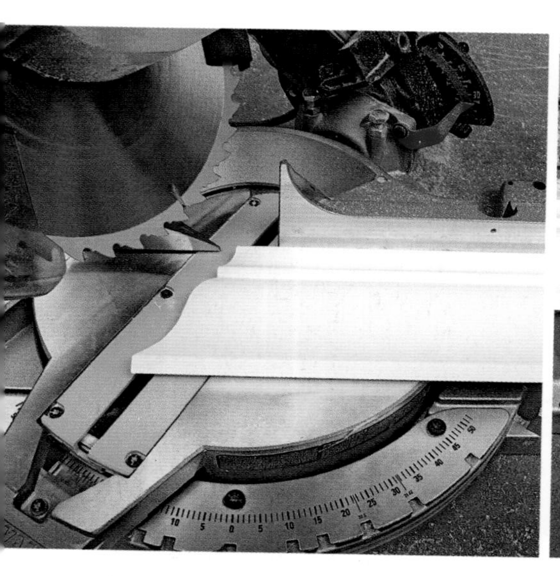

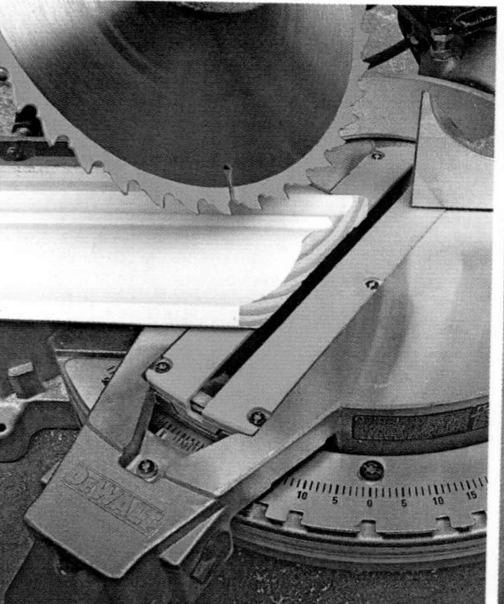

4 **CUT LEFT OUTSIDE.** To cut the *left* side of an *outside* miter, adjust the blade angle to cut to the left. Position the board on the right side of the blade, with the *bottom* edge against the fence.

5 **CUT RIGHT INSIDE.** To cut the *right* side of an *inside* miter (in preparation for cope-cutting), adjust the blade angle to cut to the left. Position the board on the left side of the blade, with the *bottom* edge against the fence.

6 **CUT LEFT INSIDE.** To cut the *left* side of an *inside* miter (in preparation for cope-cutting), adjust the blade angle to cut to the right. Position the board on the left of the blade, with the *top* edge against the fence.

Making a scarf joint

If you have a very long wall, you may not be able to buy a crown board long enough to span the distance; in that case, a good scarf joint is called for.

There's another reason for scarfing: Installing crown molding can be eased by installing two pieces rather than one along one or two walls. That way, you can start with a board that is longer than needed and fuss with a cope or miter cut on one end until you get it just right. Then hold the board in place and cut to length so it meets the other piece in the middle of the wall. If you make a neat scarf joint and sand it well, it can be nearly invisible.

1 BEVEL ENDS OF BOARDS. Cut the ends of the boards at the same bevel, around 30 degrees. (This saw has a stop at 31.62 degrees, so we went with that.) This angle helps hide the joint but also allows you to butt pieces together more easily than if they were cut at 45-degree bevels.

2 SPLICE THE SCARF. Attach one board to the wall, but hold off on nailing within a few feet of its scarf-cut end so you can adjust its position slightly. Cut the second piece's scarf joint, press it into place, and make minor adjustments as needed: You may need to move boards up or down, or shave off one end of the second piece. Once you've got a nice tight fit, apply glue to the exposed edge of the first board's scarf and nail the second board in place.

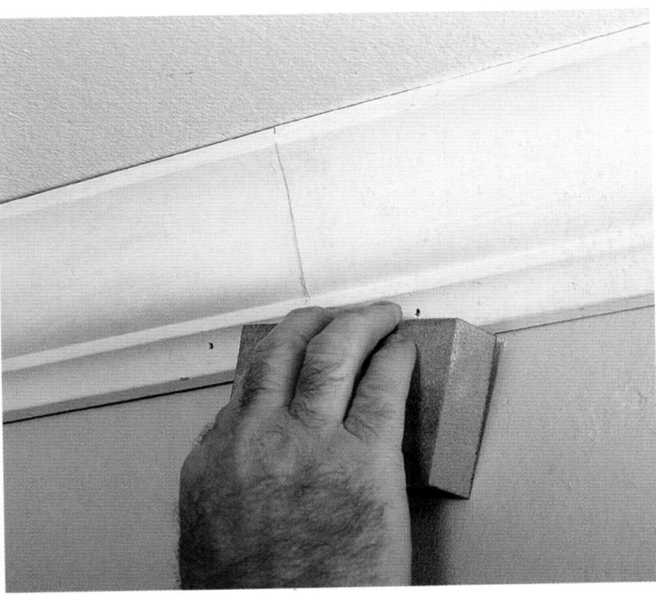

3 SAND THE SCARF. Wipe away squeezed-out glue. Allow to dry, then sand the joint thoroughly and carefully so the joint line all but disappears.

Inside Corners

If your walls and ceilings are perfectly straight, level, and plumb, you could simply cut both inside-corner pieces at 45 degrees and butt them together. But most professionals use coped joints at an inside corner. It takes more time, but you can fine-tune your joints to achieve a neat appearance even when dealing with less-than-perfect circumstances.

TIP **The end of the first (straight-cut) piece needs to be tight to the wall at the bottom, but the top will be covered so a gap there does not matter.**

1 **INSTALL THE FIRST STRAIGHT-CUT PIECE.** **Cut the first piece at a straight 90 degrees and install. Be careful to install this piece at the correct spring angle, so the flat sections on the back are snug against the wall and ceiling.**

Measuring Inside Corners

Most measurements will be from inside corner to inside corner. If you simply run a tape measure up to a corner, you won't get an accurate measurement because the tape curves as it turns the corner. Instead, measure from one wall and mark an easy-to-add-to number on the wall; here, we chose 80 in. Then measure from the other wall to your mark and add the two numbers.

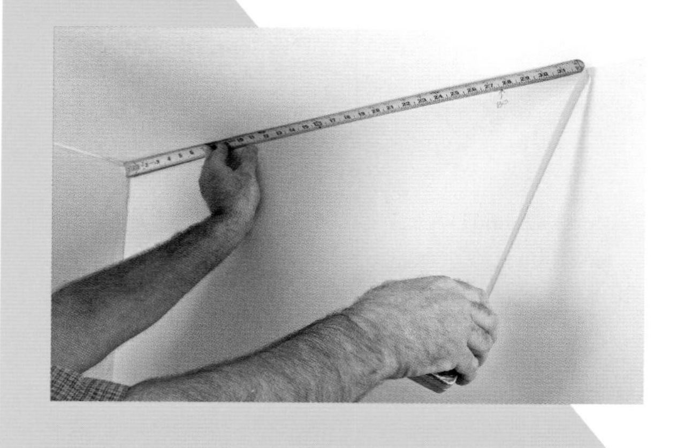

TIP **Do not drive nails closer than 2 ft. from the end of the first piece, so you can very slightly adjust its angle if needed when attaching the second piece.**

2 **MITER AND COPE THE SECOND PIECE.** **Cut the second piece for an inside miter (either left or right) as shown on pp. 158 and 160. Make a cope cut using a coping saw; be sure to back-cut, so the front edge is the longest edge at all points. See pp. 68–70 for more instructions on coping.**

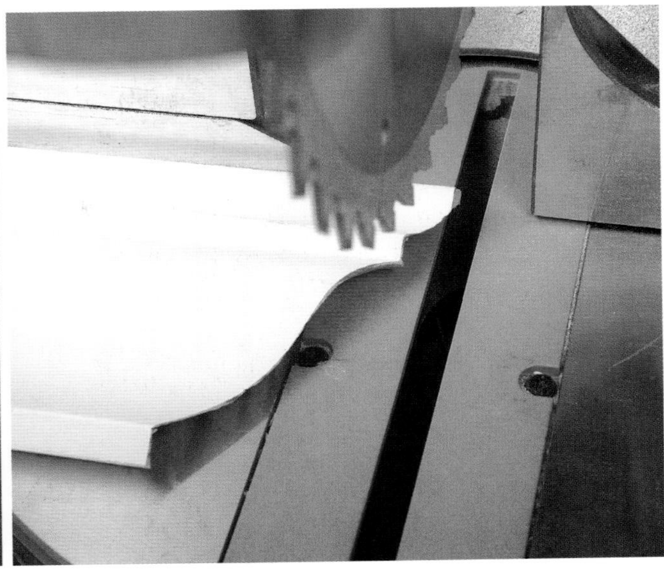

3 **NIP THE BOTTOM EDGE.** The top of the cope-cut piece will be highly visible, so cut carefully along its profile (left). The bottom edge, however, needs to be nipped off so as to expose the bottom edge of the first piece (see the photo in Step 6). Trim it with a handsaw or chopsaw (right).

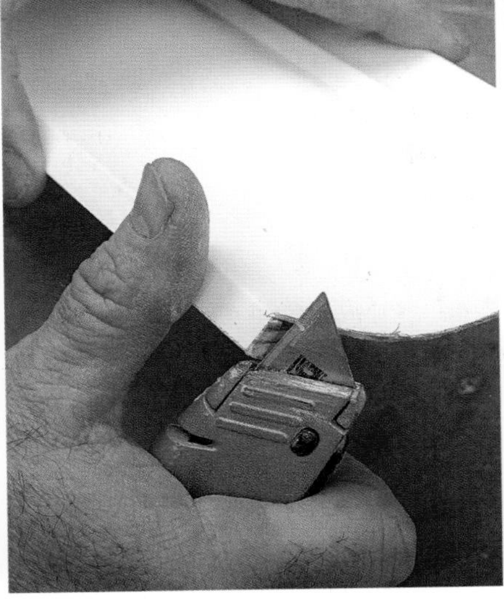

4 **MARK FOR FINE TUNING.** Hold the second piece against the first and check the fit. You will likely need to make minor adjustments. Use a pencil to mark areas that need to be shaved back.

5 **FINE-TUNE THE JOINT.** Use a knife, sanding block, or rasp to modify the second piece as needed. If the back of the piece is butting against the first, use a Surform tool, rasp, or plane to cut it back as needed. Work patiently until you get a neat-looking joint.

6 **ATTACH THE CROWN.** Snug the two pieces together; it may help to twist a bit to get things really tight. Drive nails at the bottom and top to attach. Caulk any gaps larger than 1⁄16 in.

Outside Corners

Though no time-consuming coping is needed, outside corners can be a challenge, especially if the wall corner is not plumb and square—which is often the case. Here, it's especially recommended that you start with boards longer than needed, if possible. If that's not possible, testing with scrap pieces is highly recommended.

Although an inside corner can often be successfully caulked or filled, that's more difficult for an outside corner, where you really need crisp, clean lines. So take your time, and don't hesitate to throw out a piece and start again if needed.

TIP **To miter-cut boards for an out-of-square angle, set the miter saw to one-half of the degree difference for each cut. For instance, if the angle is 92 degrees, cut each board at 91 degrees.**

1 **TEST THE ANGLE. Use a protractor-type tool like this one to test the angle of the outside corner. If it is less or greater than 90 degrees, make minor adjustments as needed.**

2 **CHECK WITH SCRAPS. Miter-cut two scrap pieces, following the instructions on pp. 158–160. Hold them in place and check the fit. Make adjustments to the scraps until you get the joint just right. Then make those same adjustments when you cut the real crown boards.**

3 **MARK FOR CUTTING TO LENGTH. Don't measure with a tape; instead, cut each board as needed for a good fit at the other end, then hold in place and mark for cutting the outside miter.**

4 **SNEAK UP ON THE CUT, SHAVE THE BACK.** Cut boards a bit long—maybe ⅛ in. or so. Test the fit and make minor adjustments as needed. With outside corners it's often the case that you need to shave the back for a tighter fit; be sure not to shave any part of the board that will show.

5 **FIRST PIECE.** Install the first piece while holding the other piece—or a scrap piece, as shown—against it. This ensures accuracy; it's easy to get it wrong if you just look at the wall's outside corner.

6 **SECOND PIECE.** Attach the second piece with nails at the top and bottom. Press it firmly into place before driving nails, so nailing does not shift it out of position. The last nails should be fairly near the edge. If there is a tiny gap, driving pin nails near the tip (inset) can sometimes help.

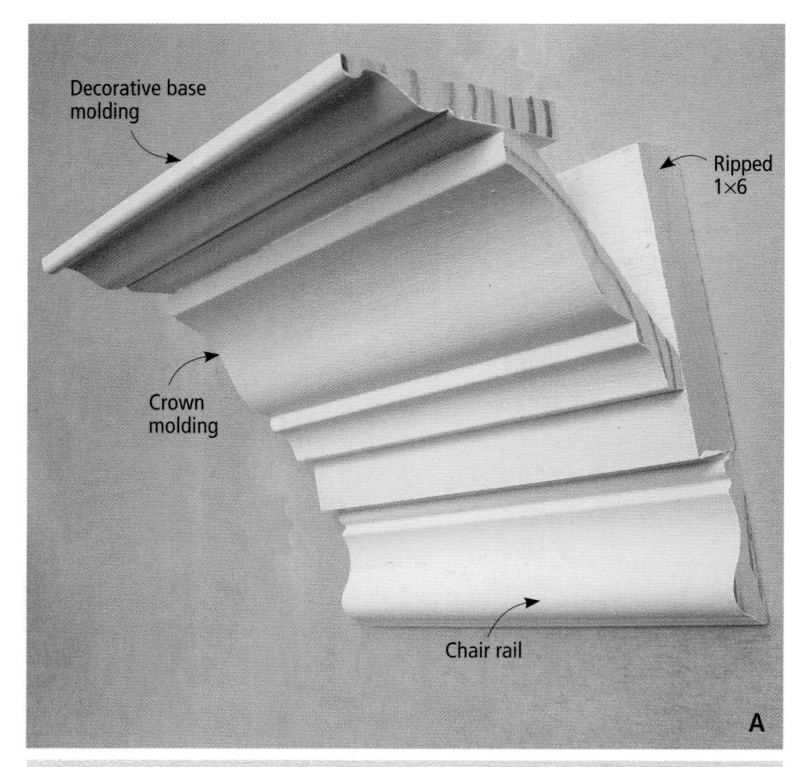

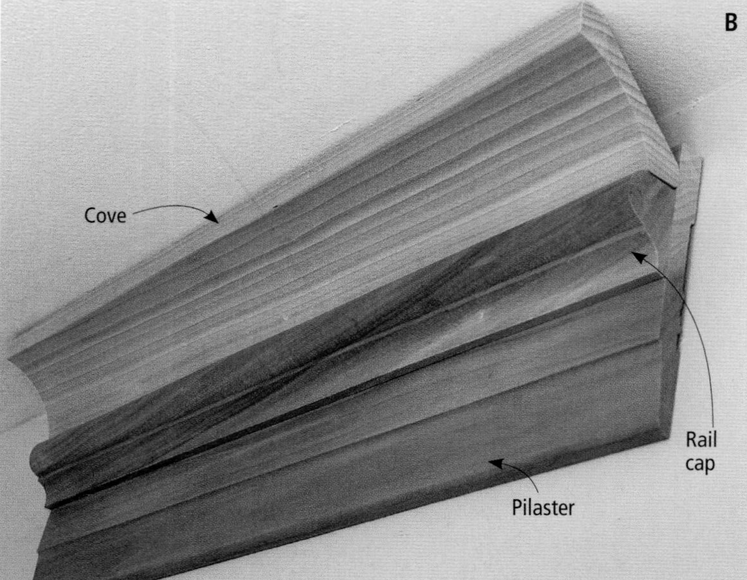

Built-Up Crown Molding

Older homes with 9-ft. or 10-ft. ceilings often have wide crown moldings, often called cornices. These are usually made with several types of moldings arranged and layered in ways both fanciful and harmonious. Ingredients in this sort of arrangement almost always include at least one crown molding, with the addition of base molding, 1× boards, cove, quarter-round or base shoe, bed molding, and perhaps even chair rail or picture molding.

Built-up crown may be as simple as a single piece of base molding at the ceiling or wall, plus crown molding. Or it may involve as many as six different pieces. It's also possible to make the wall itself part of the ensemble: If you install, say, picture molding a few inches below a crown grouping and paint the space between the same color as the molding, it gives the appearance of a very wide cornice.

Built-ups often start with a wide frieze board on the wall or ceiling. The frieze may be a plain 1×, or it could be a piece of Colonial base, whose routed front edge is visible while its wide flat width acts as a base for attaching other molding pieces.

BUILT-UP CROWN. A This built-up design is rich in appearance. Start by ripping a 1×6 to 4¾ in. wide and attach it to the wall, up against the ceiling. Attach 3¼-in. decorative base molding to the ceiling, 1½ in. out from the ripped 1×6. Install 3⅝-in. crown onto the two pieces, and add chair rail to the bottom, up against the 1×6. B This arrangement uses 3¼ in. cove, with rail cap just under it, and pilaster (often used for three-part base) under that. The result is more wide lines than details, for a sweeping look.

1. AVOID REPETITION.

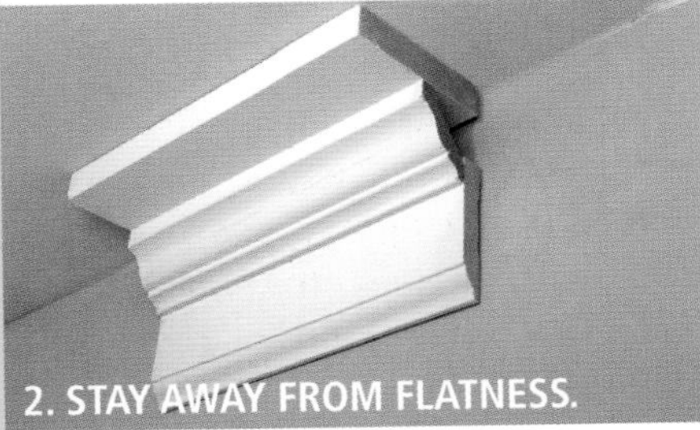
2. STAY AWAY FROM FLATNESS.

3. STRIVE FOR GOOD BALANCE.

Design Do's and Don'ts

Choose your built-up components carefully to achieve an ensemble that is both interesting and harmonious. Here are some design tips.

1. Don't use moldings with the same profile at both the wall and the ceiling; that would be too repetitious. Also, although Colonial base or other moldings with a thin edge will look good at the wall, they usually do not look good on the ceiling.

2. Avoid any flat reveal wider than 2 in., which tends to look boring. If you want a wide space, consider placing picture molding on the wall several inches below the crown molding, as shown on p. 148. Also, make the ceiling piece thick enough. Because it is not viewed along its edge, a plain, thin ceiling piece like this looks weak.

3. Here's an example of a simple but pleasing arrangement. The ceiling piece's thickness is a bit more than half the width of its reveal, and the wall piece has just an inch or so of flat reveal width and a modest reveal at the bottom.

Installing built-up crown molding

Creating a built-up crown means following the same steps as for crown molding, but multiple times. One difference: Pieces that lie flat on the ceiling or wall may be installed with miter joints rather than coped joints at inside corners. Make a sample of your design using short pieces, and be sure you will know where to put each piece, and in what order.

Also, where joists run parallel to the wall you may need to fasten a board flat to the ceiling using glue and special screws or screws with plastic anchors, because there will be no joists to nail into.

1 BUILD A MODEL. Make a partial or full model of your built-up configuration, assembling with pin nails or small finish nails. You can use this model to help position various boards on the wall or ceiling.

TIP If your walls or ceiling are wavy, built-up crown may actually be easier to install than a single piece, because the base layers can provide a flat surface for the crown molding.

2 **TEST THE CEILING PIECE.** Most built-ups include a piece on the ceiling, in this case a 1×4. Most often, only an inch or so in the front will show after the crown piece is installed. Make miter cuts at the corners. If the joint is not tight, adjust the miter saw and cut each piece a degree or less more acute or obtuse. Test and cut until the joint is tight.

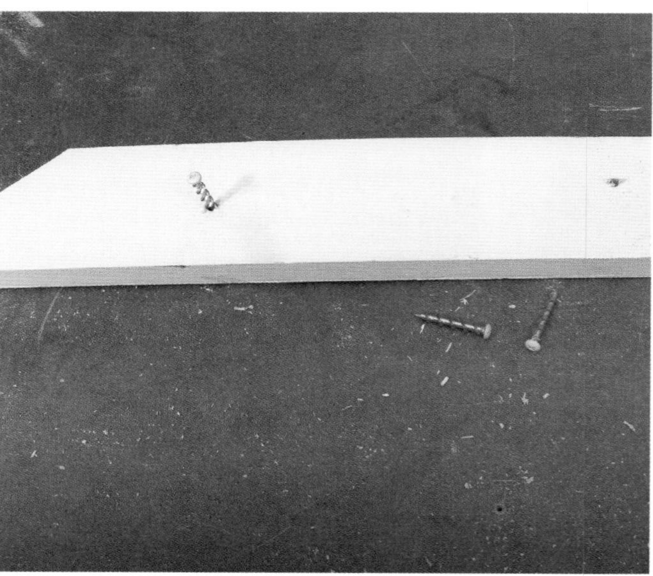

3 **ATTACH THE CEILING PIECE.** Where ceiling joists cross the ceiling piece, simply drive nails to attach. Where joists are parallel and you have no joists to attach to, you have several options. You could use toggle bolts, as shown on p. 204, or attach with screws and plastic anchors. Or as shown here, use special screws with wide, thick threads, which grab onto drywall with a fair amount of strength. In the board, drill holes that are wide enough for the screws to easily slip into, every 6 in. or so. You may not use all the holes. Apply panel or construction adhesive to the back of the board and attach with the screws.

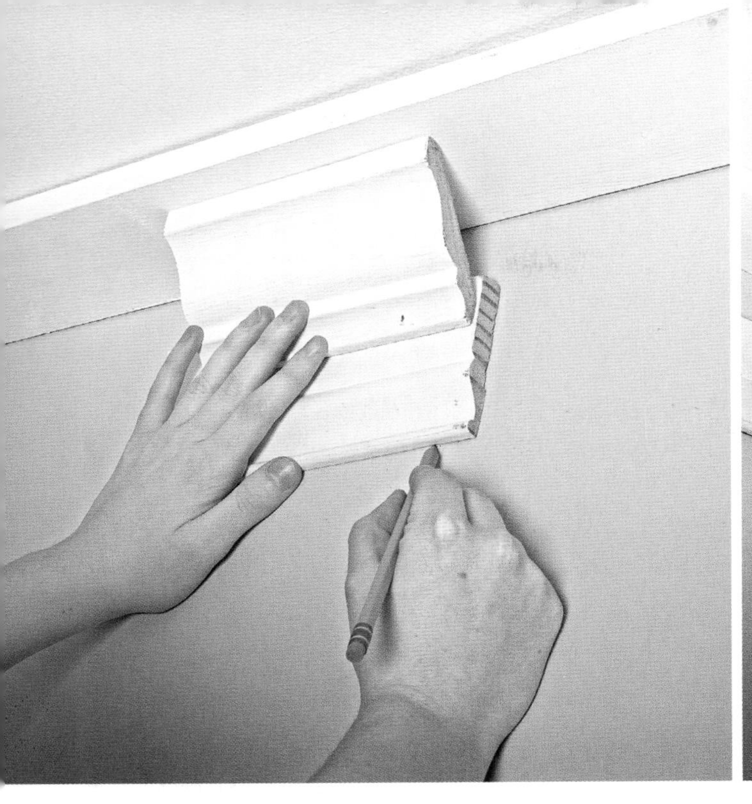

4 **DRAW A LINE ON THE WALL.** Use a model of assembled trim to mark the wall for positioning a piece that will back the crown molding. In this case, the line will be covered up when the final piece is installed. Scribe lines every few feet or so, or snap a chalkline across the wall.

5 **ATTACH THE BASE MOLDING.** Align the base molding backer piece with the scribed line and attach. At a corner you can simply miter-cut the pieces or make a coped joint. Here a piece of shim is slipped behind one of the pieces to bring its top out from the wall for a tight fit.

6 **INSTALL THE CROWN MOLDING.** At an inside corner, make a coped joint, as shown on pp. 162–163.

7 **THE FINISHED CROWN.** Install a chair rail that butts up against the baseboard molding. Fill holes and gaps with spackle or caulk, sand, and paint.

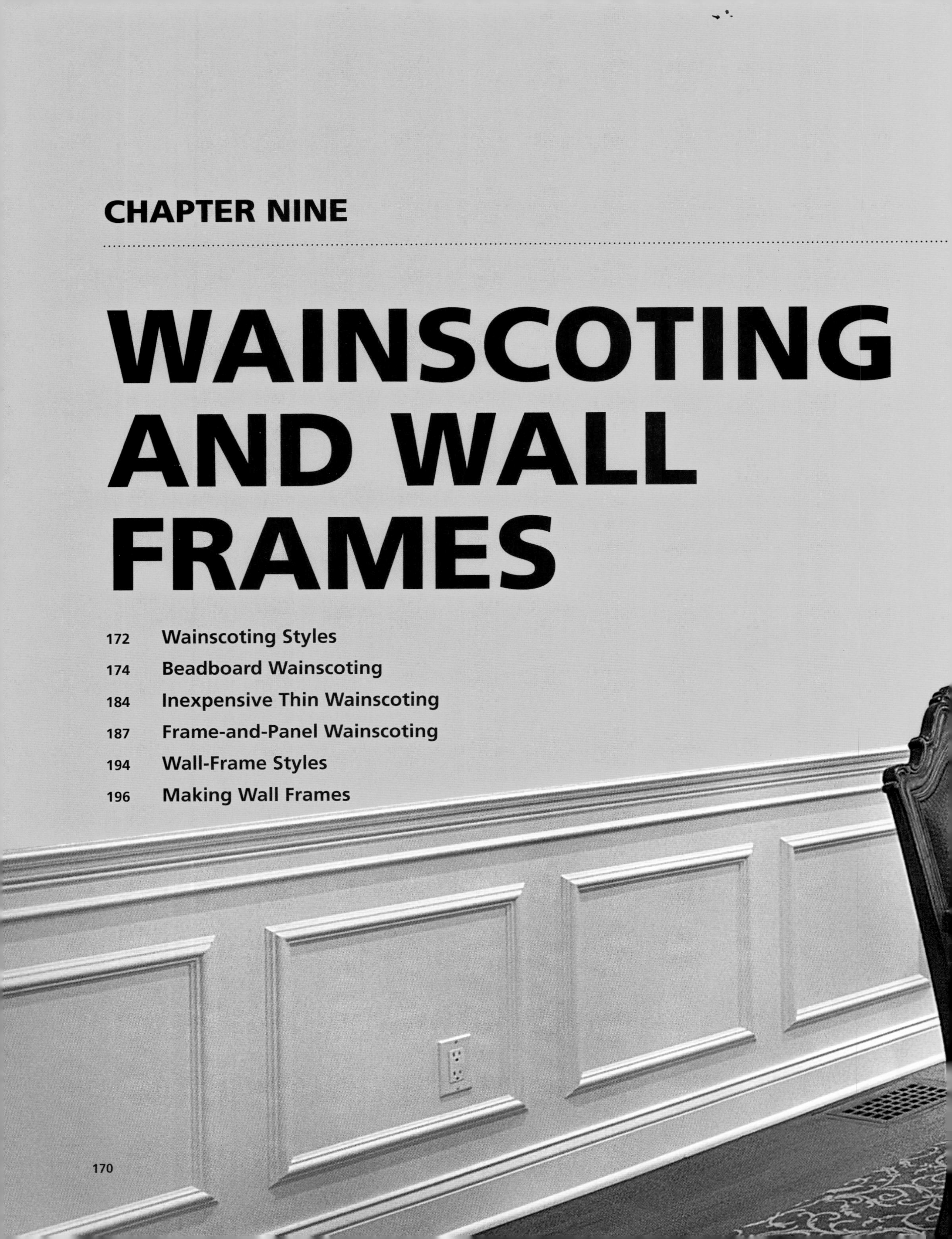

CHAPTER NINE

WAINSCOTING AND WALL FRAMES

MOST ROOMS HAVE WALLS that are essentially wide open spaces with trim only at the bottom, around windows and doors, and perhaps at the top. This chapter describes wall treatments that go beyond the framing moldings, to shape the look of the walls themselves.

Wainscoting is paneling that reaches partway up a wall. Most wainscoting is 3 ft. to 4 ft. tall, but it's not unusual for it to be 5 ft. or 6 ft. tall. Wainscoting can be made of very inexpensive painted fiberboard, of rich hardwood, or from materials that are in between in price. Beadboard wainscoting makes a room feel like a cottage. Panel wainscoting is typically composed of 1× frames laid over plywood sheeting. Though often associated with luxury, it is surprisingly easy to install and will cost less than you might expect, especially if you use materials meant for painting. Even stain-quality panel wainscoting is affordable and DIY friendly for many homeowners.

Wall frames can lend sophistication to a room and are often installed in formal areas. But they are not difficult to install and can also have a playful effect, especially if painted a bright color. They are often installed above and below a chair rail, but that's not a rule.

Wainscoting Styles

The most familiar type of wainscoting is about 3 ft. tall with vertical beadboard. But don't be afraid to try other options.

A Frame-and-panel wainscoting at a traditional height of 3 ft. playfully bounces off other rectangular shapes in this room for a satisfying, unified effect. B Inexpensive sheet beadboard is painted to match the curtains and set atop a wide baseboard, for a lovely look that belies its small expense.

C Five-foot-tall beadboard wainscoting can look great even in a room with an 8-ft. ceiling. Here, the wainscoting's cap neatly ties into the octagonal window's trim. Extra-wide baseboard molding feels at home with the arrangement. D Just for fun, try turning beadboard on its side. E Six-foot-tall frame-and-panel wainscoting is topped with a 3-in.-wide shelf. At this height you can place collectibles without fear of passersby knocking them over. F Though it looks custom made, this wainscoting can be made using standard lumber and moldings: chair rail or rail cap the top; 1x boards or strips of ¾-in. plywood for the main frame; a pair of small moldings inside the frames; and three-part base at the bottom. The large openings may be made with ¼-in. plywood, or they could be left blank, with the wall painted to mimic the look of paneling.

Beadboard Wainscoting

Beadboard may be purchased as individual pieces that fit together via tongues and grooves, or in sheets that resemble assembled tongue-and-groove pieces. The following section shows how to install ¾-in.-thick knotty pine paneling, which is 5½ in. wide. Most of the techniques for installing this type of wainscoting apply to other types as well.

Choose boards that are fairly straight along their edges (that is, free of severe bow); you will not be able to straighten them much during installation. You can do a bit more straightening of boards that are somewhat crooked (warped from end to end along their faces), but in general, the straighter the board, the better.

Understand how the boards will be fastened: In this installation all the boards will be tightly joined to the wall's baseplate, either with fasteners driven directly through the vertical boards or with a base molding that is fastened to the plate and to studs. In their middle and near the top, only every third board or so will be fastened to a stud; the others will be held in place via the tongue-and-groove arrangement and with construction adhesive that adheres them to the wall. In addition, the very top of the boards may be held with a top cap and a piece of molding under it.

If your floor or walls are out of level or plumb, resist the temptation to follow those lines and install the pieces out of plumb; that always leads to installation problems eventually. Install the first boards on a wall nice and plumb, and check every few boards to be sure they are still plumb.

TIP **For a rock-solid installation, remove a 4-in.-wide horizontal strip of drywall near the top of where the wainscoting will go and replace it with ½-in. plywood. That will provide you with a firm surface for nailing all the pieces. However, the method shown here is easier and will hold well, as long as the boards are not badly warped.**

1 **REMOVE THE APRON. If the wainscoting will bump into trim or other obstacles, you may choose to remove it or cut around it. Usually it looks neater if you remove a molding such as window apron (as shown), and then reattach it after the wainscoting is installed.**

2 **CUT THE BOARDS. Determine the height for the wainscoting and factor in the top cap and the base molding if you will make rabbeted base, as shown in the sidebar on the following page. Gang-cut the tongue-and-groove boards to the same length.**

3 **MARK THE STUDS. Draw a level line on the wall about ⅜ in. above where the top of the paneling boards will be, and mark where the studs fall. (This line will get covered when you install the cap.)**

1A. CUT THE RABBET IN THE BASE.

1B.

2A. NOTCH FOR INSIDE CORNER.

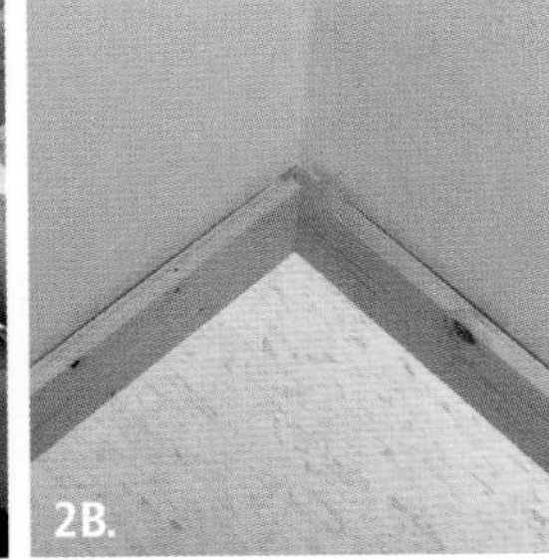

2B.

3. INSTALL THE WAINSCOTING.

Rabbeted Base Molding

A rabbeted base makes for a sleeker, less rustic appearance, because it appears to be only $\frac{3}{8}$ in. rather than $\frac{3}{4}$ in. thick. To make the base, rip-cut a long piece of the paneling to remove the tongue and the groove, or use a 1×4 with a similar knotty appearance.

1. To make the rabbet, adjust a tablesaw's blade to cut $\frac{3}{8}$ in. deep, and position the fence to make a cut $\frac{3}{8}$ in. thick (**1A**). Rip one direction, then the other to create the rabbet (**1B**).

2. Make a small cut at the end of one piece (**2A**) so when two pieces are put together there will be space for paneling boards in both directions (**2B**).

3. Assemble and attach the base and the paneling boards at the same time. Check that the base is level; set it on shims as needed. Slip several pieces of the paneling into the rabbet, and use a scrap piece to hold the base away from the wall as you drive nails through the base.

4 **MEASURE THE BOARDS.** Plan the installation so you don't end up with a narrow sliver at one end. Place several pieces against a wall and measure to see how much horizontal space each takes up. (It will probably be close to 5½ in.) Measure the entire length of the wall, and divide by the horizontal dimension you found. If there is a remainder of 2 in. or less, rip-cut and install a partial-width piece at the beginning of the wall so the partial piece at the other end will be roughly the same width.

TIP Many carpenters prefer to install paneling boards with the tongue facing out, so the next board gets installed by slipping its grooved end onto the tongue. Others prefer to install groove-side out. Either way works, but experiment with pieces to see if one method is easier than another.

Top Guide

If you do not have a rabbeted base, you can attach a temporary guide to keep the tops of the paneling boards level. This will speed up the installation. Use any straight board. Check it for level, and drive nails or screws into studs to attach.

5 **MARK THE FIRST PIECE.** If you start installing at an inside corner (which is usually the case), check each wall for plumb. Start installing onto the wall that is the closest to plumb. Usually the side of this piece will be covered by the thickness of the piece installed on the other wall. But if the other wall is significantly out of plumb, hold the piece plumb and mark for cutting as shown, with a straight board. This board does not have to be rip-cut precisely.

TIP Assembling tongues and grooves is often easily done, but if you're having trouble getting a board to slip into place, pull it away and check its tongue or groove—and the tongue or groove of the previous piece—for dents, splinters, and other defects that are getting in the way. Slice or pry them out of the way as needed.

6 RIP-CUT THE FIRST PIECE. If you need to make a rip cut that is not parallel with a board's edge, mark it for an angled cut and cut using a circular saw. Have a helper hold the board as you cut it, or you may be able to clamp to a straightedge guide, as shown.

7 SCRIBE AT THE CORNER. Attach the first inside-corner piece. The piece on the other wall should fit tightly against it. Hold it in place so it is plumb and scribe a cutline. Usually there will not be a great difference in width between the top and bottom, so you can scribe by holding a pencil flat against the first piece, as shown. If the gap is wider, use a compass.

TIP You may choose to cover the inside corner with a trim piece, such as quarter round or cove. In that case you can skip Steps 7 and 8.

8 FINE-TUNE THE CUT. If necessary, rip-cut the second corner piece as shown in Step 5; take care not to cut too narrow, but it's OK to be a little wide. Then use a belt sander equipped with a 50-grit belt to fine-tune the cut.

9 NAIL THE BOARD PLUMB. Install the first piece on a wall by checking for plumb, then driving nails or screws into the bottom plate and a stud. Usually there is a stud near the corner. If there is no stud, first apply adhesive to the wall (next step) and drive two nails or screws into the drywall, to keep the board from moving out of plumb.

10 APPLY ADHESIVE AS NECESSARY. Where no stud is available for attaching boards, apply squiggles of panel or construction adhesive.

11 TAP INTO POSITION. Slip two boards into the tongue or groove of the previous boards. Check that the joints are all the same width. If needed, slip a scrap piece of paneling with an exposed tongue or groove onto a board's side and tap with a hammer to bring the boards into alignment.

12 NAIL AT THE STUDS. When you come to a stud (about every three boards), check the board for plumb. If the board is out of plumb, gently pry out the top or bottom. Drive finishing nails or screws to attach the board.

Cutting Around a Receptacle

1A. MARK FOR THE CUTOUT.

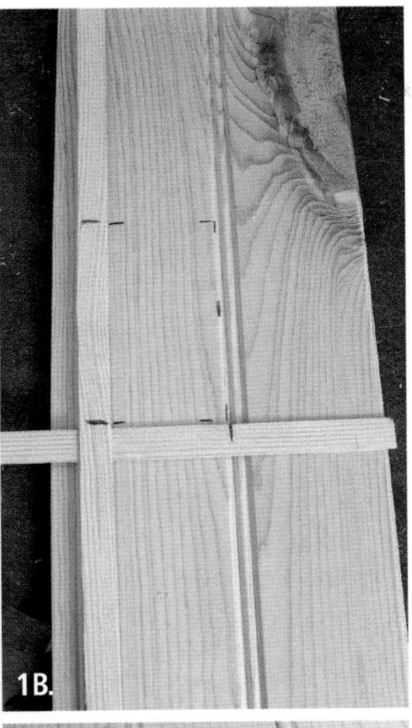
1B.

2. MAKE THE CUTOUT.

3. ACCOUNT FOR PANEL THICKNESS.

4. INSTALL THE COLLAR.

5. ATTACH THE RECEPTACLE.

At a receptacle, you'll need to cut a rectangular hole that is at least as large as the electrical box, but small enough so it is covered by the receptacle's cover plate. In particular, the hole should be tall enough so you can drive the receptacle's mounting screws at the top and bottom, but short enough so the receptacle's "ears" can rest against the wood. You have some wiggle room but not a lot, so measure carefully but don't feel you have to be precise.

Before you touch the receptacle, shut off power to the receptacle at the service panel and test to be sure power is off in both of the outlets.

1. Install the last board possible up to the receptacle. After you are sure the power is off, unscrew the mounting screws and gently pull the receptacle out, taking care not to disconnect any wires. Measure over from it and up from the floor to find the four sides of the cutout. You can use a tape measure or you may find it simpler to use small slats, as shown here (**1A**). Transfer the measurements to a panel board (**1B**).

2. Use a square to mark the outline of the cutout. Drill a starter hole inside the cutout, and cut the hole with a jigsaw.

3. Because the paneling will move the receptacle out ¾ in., its mounting screws will probably not reach when you reinstall it. Buy 6-32 screws that are longer than the originals. For safety and electrical code compliance, install a metal receptacle box extension collar that will bring the box up flush to the face of the installed board.

4. Slip the metal collar inside the electrical box.

5. Push the receptacle into place and drive the long mounting screws so its ears snug up against the wainscoting.

13 **RIP THE CAP.** The wainscoting's cap is often made of knot-free wood, so it is easier to clean. A good width for a cap is achieved by ripping a piece of 1×4 in half to make pieces that are about 1⅝ in. wide.

14 **ROUND THE CAP EDGE.** The cap can be left with a squared front edge, or you may choose to soften the edge with a router equipped with a round-over bit.

15 **MARK AT THE CORNER.** You could simply cut the cap at 45-degree miters, but corners are rarely perfectly square. Place one of the pieces in position and mark its back at the corner (left). Then lay the other piece on top, and mark where they will meet in front (right).

TIP The top cap shown in the photos is narrow, so it doesn't become a bumping hazard. You may choose to make at least some of your top cap wide enough to act as a shelf.

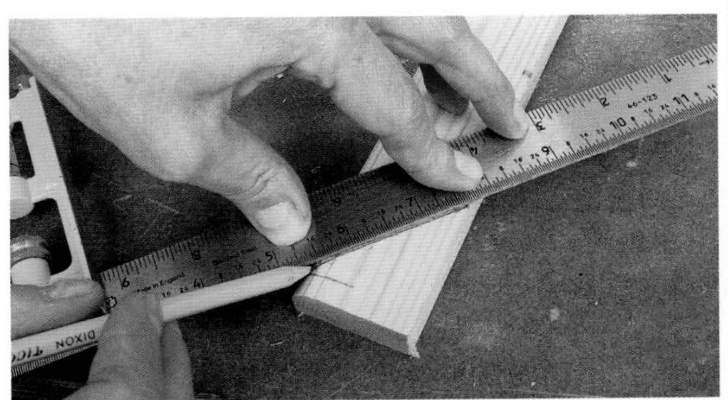

16 **CUT THE MITER.** Join the two marks by drawing a line with a straightedge (top). Cut along the lines with a miter saw (above).

17 **NAIL THE CAP CORNER.** Check the fit at the corner. If you need to change the angle slightly, make the adjustment on both pieces rather than just one; otherwise, the cuts will be different lengths. Attach the top cap to the top edges of the panel boards, and drive a nail through each side at an outside corner.

18 **SCRIBE AS NEEDED.** Where a wall waves a bit, scribe a line with a pencil. Remove material with a belt sander so the piece snugs against the wall.

19 **CAP-TO-CASING DETAIL.** Where a cap meets casing or other molding, make a cut at about 30 degrees that dives into the edge of the molding, for a neat appearance.

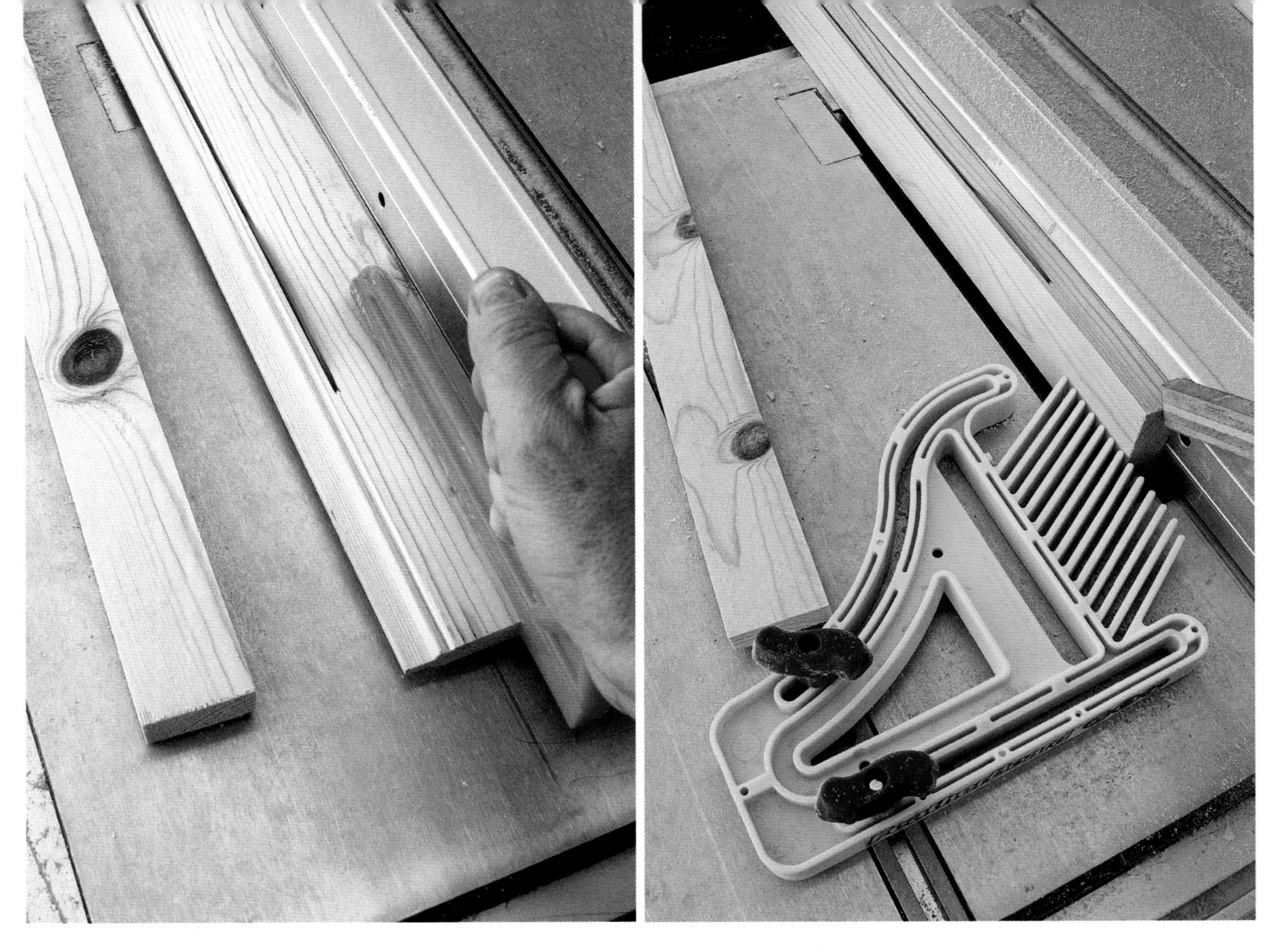

20 **RIP-CUT THE OUTSIDE CORNER.** To trim an outside corner with a cottage look that is not too thick, first rip-cut pieces of paneling. Rip one piece at 2 in. wide and another at 2½ in. wide (left). Then turn the pieces on edge, adjust the fence, and rip-cut these pieces to a thickness of ½ in. (right).

Using an Angle Tool

You can use a protractor-type angle tool to measure an inside or outside corner's angle. If you find a corner is out of square, make miter cuts that are half as wide or narrow as the corner's deviance from 90 degrees. (For instance, if the angle is 92 degrees, cut both pieces at 91 degrees.)

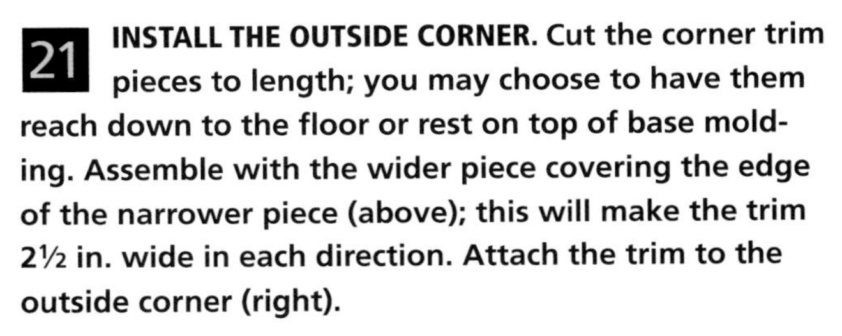

21 **INSTALL THE OUTSIDE CORNER.** Cut the corner trim pieces to length; you may choose to have them reach down to the floor or rest on top of base molding. Assemble with the wider piece covering the edge of the narrower piece (above); this will make the trim 2½ in. wide in each direction. Attach the trim to the outside corner (right).

Rabbeted Outside Corner

1. CUT A SMALL RABBET.

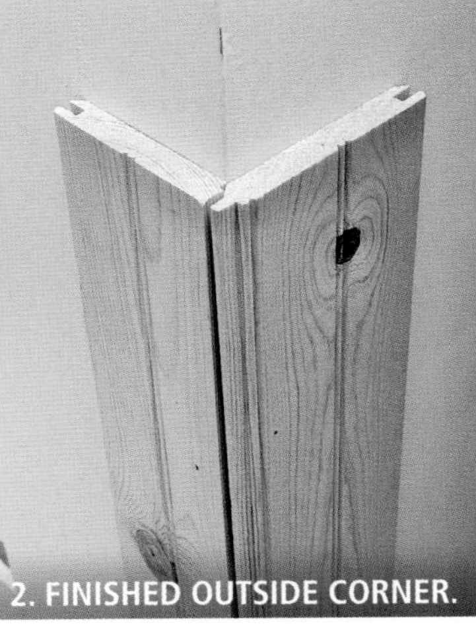

2. FINISHED OUTSIDE CORNER.

If an outside corner is nice and straight, you may choose this detail instead of corner molding. It is easiest to install if you start at the outside corner and then install the other boards from there.

1. Rip-cut the tongue off of a paneling piece. Then rip a small rabbet along the back edge, about 3⁄16 in. deep and 3⁄16 in. wide.

2. When installed adjacent to a board with its tongue showing, it produces a pleasing beaded effect.

Inexpensive Thin Wainscoting

At a home center you will find ¼-in.-thick beadboard wainscoting, either in sheets or in strips about 7 in. wide and most often precut to length. One type is made of unpainted pine that splinters easily, whereas another type is low-density fiberboard that can be broken off with your hand. It is not what you'd call a quality product, but if installed firmly onto a wall and kept protected with paint or finish, its appearance can be as rich as real wood paneling, and it will last for a very long time. In fact, you may prefer its closely spaced shallow grooves to the look of ¾-in.-thick paneling.

Most of the installation instructions are the same as for thicker paneling. Here we hit some of the high points and some of the material's quirks.

THREE PIECES. This type of beadboard is sold in sheets or strips, along with a compatible top cap, at left, and a notched baseboard, at right.

TIP **When using this thin material, don't attempt to make inside or outside corners without trim. It's difficult to get a tight fit, and even if you do, the joint is likely to come loose in time when the room's humidity changes.**

1 CUT BY HAND. Because the base and cap are thin and soft, it's easy to cut them with a hand miter saw. Of course, you can also cut with a power miter saw.

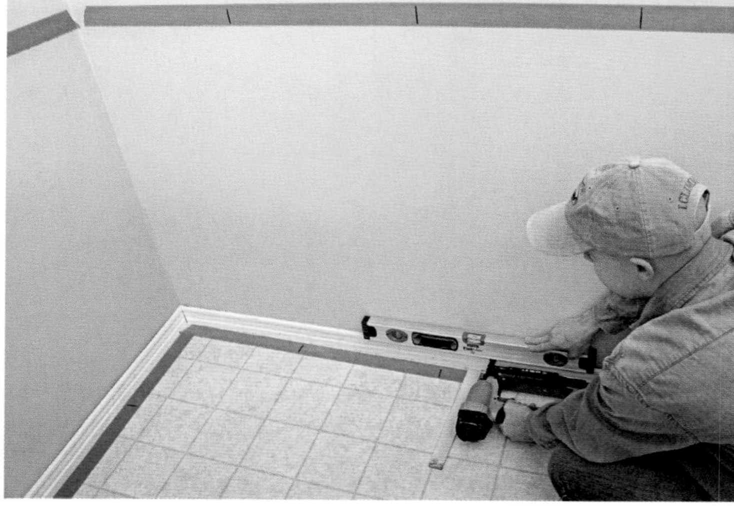

2 ATTACH THE BASE. Use painter's tape and a felt-tip marker to mark locations of studs on the wall just above where the top cap will go, and on the floor. Cut the base molding to fit, with 45-degree bevels at the corners. Set base pieces on the floor and use shims to bring them level. Make sure abutting base pieces are at the same height.

3 **FIT THE INSIDE CORNER.** Check and perhaps adjust the layout to be sure you will not end up with a narrow sliver at one end. At an inside corner, butt two strips together and check for plumb (left). The joint will be covered with molding later (Step 8), but if a gap is larger than ¼ in., scribe a cutline and use a plane, a Surform tool, or a power saw to cut as needed (right).

TIP Work to achieve accurate plumb and level when you install the base and the first panel strips. Any minor imperfection will grow worse as you continue to install more strips.

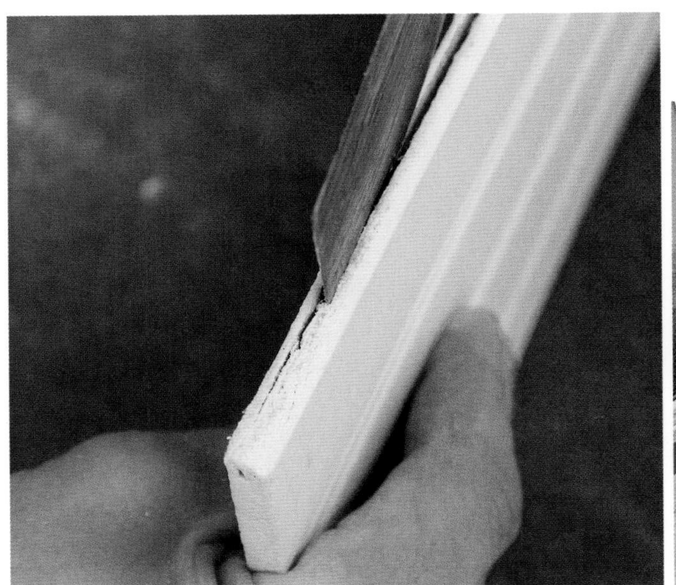

4 **OPEN UP THE GROOVE.** Before you apply any adhesive to a strip, check that it will fit. Usually this works easily, but it's not unusual for the tongue or groove to have flaws that make it difficult or even impossible to fit tongue to groove. Here you see a common flaw—too much paint in the groove. Use a putty knife, chisel, or handsaw to open the groove. Work carefully, to avoid breaking out the front of the strip.

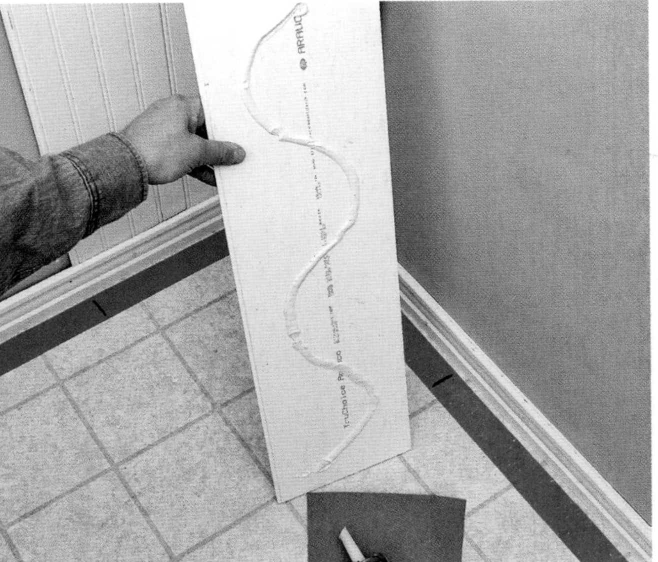

5 **APPLY ADHESIVE.** You can usually nail the first piece into a stud. For the next pieces, apply a squiggle of construction or panel adhesive to the back. Slip each piece down into the base's rabbet, then press into the wall as you slide it over for a tongue-and-groove fit.

6 **TAP OVER.** Check each strip to be sure it is fully slid into place. Make a block with an exposed tongue or groove to use as a tapping tool. If a strip needs persuading, slip the tapping tool onto it and hold a scrap of wood against it as you tap the strip over.

7 **NAIL THE STRIPS.** Wherever you meet a stud, drive a finish nail. If any strips are not lying tight against the wall, drive angled nails into drywall only to snug them up.

8 **ATTACH CORNER MOLDING.** Cut the top cap at 45-degree bevels like the base, and drive nails to attach it to the wall. At an inside corner, attach a modest molding such as ½-in. cove or quarter round. At an outside corner, attach outside corner molding.

9 **CAULK THE TOP EDGE.** Where the cap meets the wall, apply a bead of caulk and wipe with a damp rag or your finger to fill the joint neatly. You may need to touch up the wall with a bit of paint.

Frame-and-Panel Wainscoting

Frame-and-panel wainscoting is surprisingly easy to build and has a professional, finished appearance. Build a simple frame of 1× lumber, attach sheets of plywood behind it, and attach to the wall. Add a top cap and perhaps base shoe, and you've turned a plain room into something out of a British costume drama.

TIP **If you want to stain rather than paint the paneling, select the plywood carefully. Wide, wavy grain looks OK stained, but tightly vertical grain looks much better. You might choose, for instance, oak plywood described as "plainsawn" or "quartersawn."**

There are a few design details to keep in mind:

- As much as possible, work to keep the vertical frame pieces evenly spaced.
- At inside and outside corners, plan for vertical frame pieces that look equal in width; to achieve this, one of the pieces must be 3/4 in. wider than the rest.
- Plan so receptacles or other obstacles will fall entirely in a panel or entirely in a frame piece.

TIP **If possible, plan so that full 4-ft.-wide plywood panels can be used, with the seams between them falling somewhere in the middle of the frames. However, if doing this will throw off your desired layout, you may need to cut pieces of plywood to width to fit.**

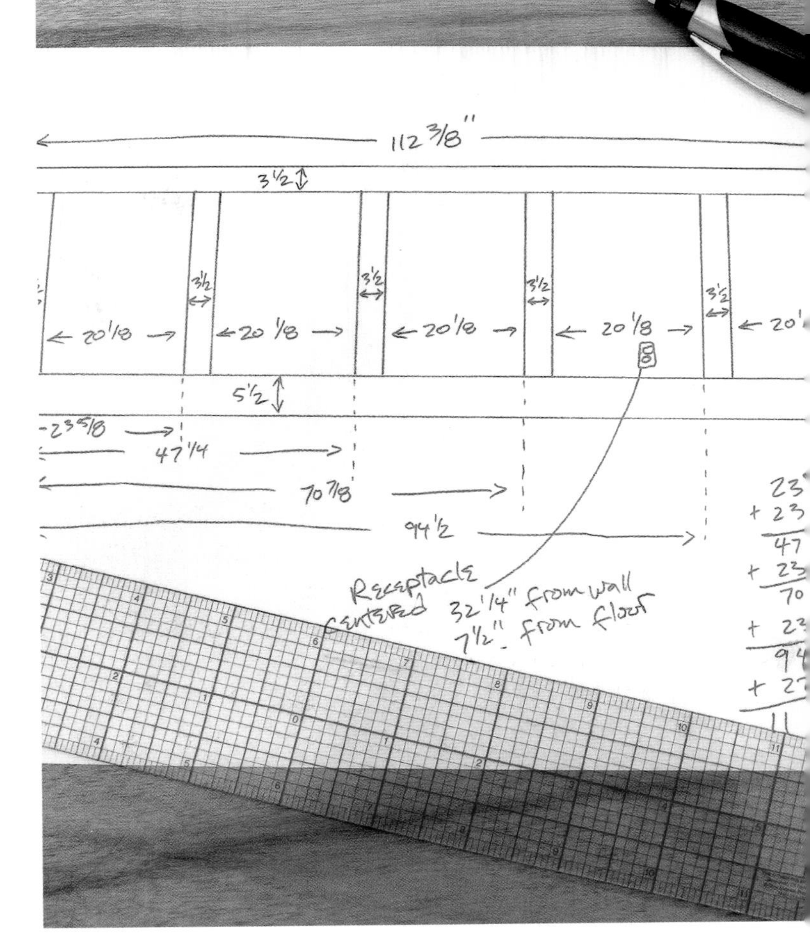

1 **DRAW IT OUT.** **Make a scale drawing of the frame for each wall, with evenly spaced verticals. Make sure receptacles do not overlap from panel to frame piece.**

2 **CUT THE FRAME PIECES.** **In the plan for this wainscoting, the top horizontal and all the verticals are 1×4s, whereas the bottom piece is a 1×6.**

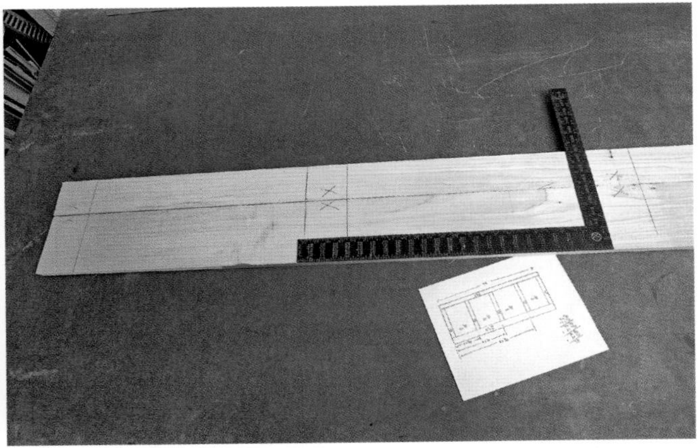

3 **LAY OUT THE HORIZONTALS.** **Lay the horizontal frame boards side by side, with the good sides facing down. Working from your plan, mark the positions of the vertical pieces.**

4 **POCKET-SCREW THE FRAME.** Set up a pocket-screw jig and its drill bit for ¾-in. material (see p. 78) and drill for two pocket screws in the back of each end of the vertical frame pieces (left). Position the verticals against the layout marks and drive 1¼-in. pocket screws to attach (right). Check for square as you work.

TIP The frame could also be joined with biscuits instead of pocket screws. See p. 74.

5 **INSTALL THE PLYWOOD BACK.** Cut pieces of plywood to fit: The size does not have to be exact, as long as they fully cover the spaces between the frame pieces and lap onto the frame pieces enough for driving fasteners. Place the plywood pieces with the good side facing down. Drive ¾-in. staples, nails, or small screws every 6 in. or so into the frame pieces.

6 **SAND AS NEEDED.** Turn the panel over and check for any imperfections. If there is a gap between plywood and frame, drive more staples to pull it tight. If a frame piece is a bit proud of an adjacent piece, sand with a random-orbit sander to even the surfaces, then hand-sand along the grain.

Cutting a Hole for a Receptacle

When you come to a receptacle, cut a hole just large enough for reattaching the receptacle, but small enough to be covered by the cover plate.

Shut off power at the service panel, and test to be sure power is off. Remove the cover plate, unscrew the receptacle mounting screws, and pull the receptacle gently out, leaving it attached to the wires.

1. MARK FOR THE RECEPTACLE HOLE.

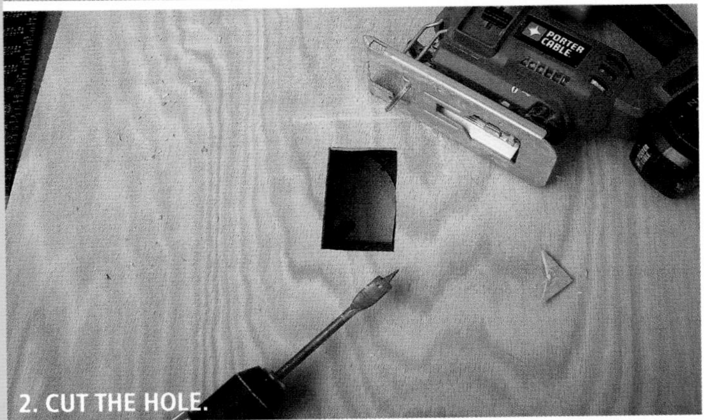

2. CUT THE HOLE.

3. ATTACH THE RECEPTACLE.

1. To mark for the receptacle hole, measure over from a nearby wall for the two sides of the box, and up from the floor for the top and bottom. Mark the panel with those dimensions.

2. Drill a starter hole, then cut with a jigsaw.

3. Because the paneling is only 1/4 in. thick, you may be able to reuse the original mounting screws to attach the receptacle, or you may need to buy slightly longer ones.

7 **SHIM AT INSIDE CORNERS.** Set two pieces together at an inside corner, press them against the walls, and check for plumb and a tight fit. If the top or bottom of the first piece (the one with the wider vertical piece) needs to be moved out from the wall, cut a piece of shim so it is the correct thickness to make up the gap (left). Attach one or more shims of this thickness to the wall (right).

TIP At an inside corner, one of the vertical pieces should be ¾ in. to 1 in. wider than the other verticals. That way, when assembled, both sides will appear to be the same width.

8 **SHIM AT THE FLOOR.** Shim panels as needed to make them the same height, as well as perfectly level.

Attaching a Panel with Plugged Screws

Attaching a panel with plugged screws adds a woodworker touch in keeping with the spirit of paneled wainscoting. It takes a bit more time but breathes a noticeable classiness into a room.

1. Buy a plug-cutting bit and a pilot/counterbore bit in matching sizes.

2. Cut a series of plugs out of a scrap piece of the panel material.

3. Drill a pilot-counterbore hole (**3A**), then drive a screw to fasten the panel (**3B**).

4. Drip a bit of glue onto a plug, press it into the hole, and tap with a hammer.

5. Cut the plug nearly flush with a chisel or fine-toothed saw.

6. Once the glue has dried, sand smooth.

1. DRILL BITS.

2. MAKE THE PLUGS.

3A. DRILL AND DRIVE.

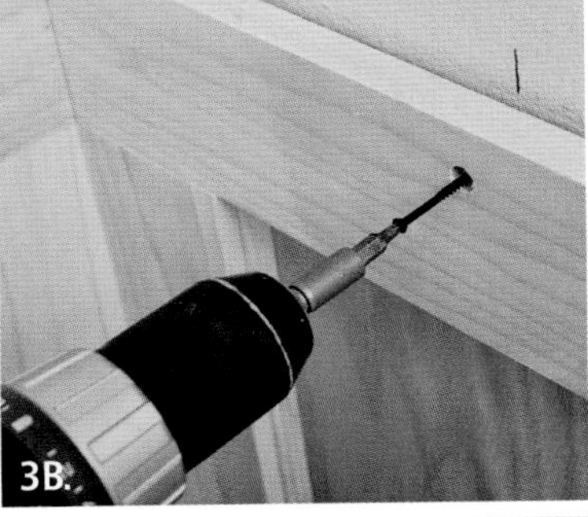

3B.

4. TAP THE PLUG.

5. CHISEL OFF THE EXCESS.

6. SAND SMOOTH.

9 **SCRIBE AND SAND.** Attach the first panel and press the second one against it. If they do not meet neatly and tightly, hold a pencil flat against the first panel to scribe a line on the second (left). Use a belt sander with a 50-grit belt to remove material and create a line that is parallel to the scribe line (right).

10 **NAIL THE PANEL.** Once panels are well aligned, attach by driving finish nails into studs. Or screw and plug, as shown in the sidebar on p. 191.

11 **FIT THE OUTSIDE CORNER.** At an outside corner, one of the panels should have a vertical piece that is ¾ in. narrower than the other. Hold the panels in place and scribe a line if they do not meet perfectly. Cut or sand along or parallel to the scribed line, attach the two pieces, and sand smooth.

Decorative Options

Simple 1× frame pieces make for crisp, neat lines, but you may prefer something a bit more dressed up. One way is to rout a profile along the inside edges of the frame before attaching the plywood. The other method is to add small trim boards after the plywood has been attached.

1. To rout the edges, use a self-guiding router bit and hold the router flat on the boards as you cut. Practice on scrap pieces until you are certain of your skills. This method produces a rounded detail at the corners.

2. To add trim, cut pieces of ½-in. cove (**2A**) or fluted quarter-round (**2B**) at 45-degree miters and attach them with glue and pin nails.

1. ROUT THE EDGES.

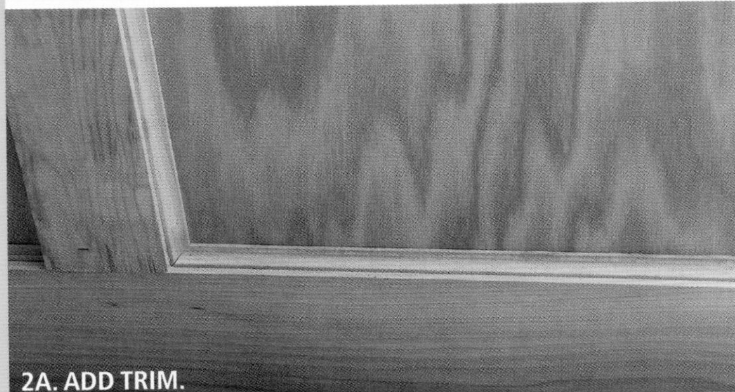

2A. ADD TRIM.

2B.

Wall-Frame Styles

Wall frames are usually made of narrow trim, such as 1⅜-in.-wide base cap (the type that does not have a notch, so it rests entirely against the wall). Wall frames can add elegance and a sense of order to a room. Shown here are some style options.

A

B

C

A Above and below the chair rail, wall frames of two different widths add a playful touch. B Wall frames march along the lower wall between base molding and chair rail. The entire area is painted white, giving the impression of frame-and-panel wainscoting. C In an informal place like a child's bedroom, simple wide lattice can artfully and casually divide up the walls. D Wall frames are not just for great rooms. Here, they help make a mudroom an inviting entry.

Making Wall Frames

Make scale drawings of your walls and plan for rectangles that are evenly spaced along the length of the wall and evenly spaced between base molding and chair rail or crown molding. Also make sure the frames will not bump into receptacles and other obstacles.

Wall frames are often installed in conjunction with chair rail. Horizontal frames are usually placed below the chair rail, with vertical frames above; in this arrangement, the frames above and below are the same width and are stacked in vertical alignment. For a room with an 8-ft. ceiling, chair rail is often placed 36 in. to 42 in. above the floor; rooms with 9-ft. or 10-ft. ceilings may have chair rail 60 in. above the floor.

The spaces between wall frames and other trim elements (such as other wall frames, baseboard, chair rail, and other moldings) usually look best at about 3 in. Not all frames need to be the same width: You may choose to alternate two different sizes, or have the frames on each end of a wall be narrower than the rest.

To plan spacing and size, follow these steps:

- Measure the wall's width and tentatively make a guess at the number of frames you want.
- There will be one more gap than the number of frames. Multiply this number by 3 in., or whatever gap size you've chosen. Subtract this number from the wall's width.
- Divide the resulting number by the number of frames. This gives you the width of each frame.
- Make a drawing using these dimensions. If you don't like the way it looks, try doing the calculations with one fewer or one more frame.

Making and installing frames

It is possible to draw a grid of layout lines on the wall and then cut individual pieces and assemble them on the wall. However, it's a bit easier to assemble the frames before attaching them to the wall, even though they will be pretty fragile until they are fastened in place.

TIP **When mapping wall frames, some designers try to adhere to the "golden rectangle" principle, which holds that the most pleasing rectangle is one in a proportion of 5 to 8. For example, rectangles that are 15 in. by 24 in., 25 in. by 40 in., or 35 in. by 56 in. would qualify as golden. Go for the gold if you can, but often wall sizes do not allow for it.**

1 MARK THE LAYOUT. **Draw some layout lines, or at least make some layout marks, on the wall. Usually you can use spacer pieces to keep the frames aligned for height, so you don't need horizontal lines. However, it's good to have vertical lines or marks to make sure the frames will end up with even spacing widthwise. Hold 3-in. spacers (or wider or narrower spacers, if you choose) and measure to make sure of the frames' heights. Also double-check the width measurements.**

2 MAKE AN ASSEMBLY JIG. **Make a very simple jig for assembling the frames by attaching a piece of plywood, with two adjacent factory edges, onto a larger plywood sheet. A hold-down clamp like the one shown is helpful.**

3 **CUT THE FRAME PIECES.** Once you're certain of the frame sizes, cutting is easy: Cut each piece with a 45-degree miter at each side. Each piece, from tip to tip, is the dimension of the width or height of the frame. To keep all the pieces exactly the same size, set up a simple stop, as shown, rather than measuring each piece.

4 **ASSEMBLE THE FRAME.** Apply a dab of glue to one of the two pieces to be fastened together. Hold the pieces firmly in place; clamp one of them with a hold-down clamp or a regular squeeze clamp. Examine the joint carefully to make sure it's nice and tight. Drive nails with an air nailer, or drill pilot holes and drive trimhead screws to fasten.

5 **APPLY ADHESIVE.** Carefully carry the frame to the wall and test the fit. Remove, and apply a thin bead of construction or panel adhesive to the back.

TIP For 1⅜-in.-wide base cap, as shown, use 1¼-in. finish nails, one in each direction. If you don't have an air nailer, you will find it easier to drive screws than hand nailing, which often knocks the boards out of alignment as you pound.

6 **NAIL THE TOP OF THE FRAME.** Align the top piece against the side layout marks and a spacer for height. Drive finishing nails at angles into the drywall to hold it in place until the adhesive sets.

7 **PLUMB THE SIDES AND NAIL.** Check the sides for plumb and drive more nails into the sides and the bottom.

CHAPTER TEN

BEAMS AND MANTELS

WE'LL END THIS BOOK with two larger-scale projects that will dress up a room impressively, but that are actually not difficult to build. Beams add a traditional touch and transform a ceiling from a blank sheet to a point of interest. Mantels and fireplace surrounds create a focal point for a living room or great room. They can be installed onto a traditional brick fireplace or a modern gas-fueled firebox. The design we show in this chapter has distinctive features, but its basic lines can be layered with decorative flourishes to create most any look. Trimwork does not call for special talents; the skills you need are taught in this book. However, every joint needs to be tight, and you must work carefully and systematically. If you have a devotion to detail and the patience to treat every board with care, you can achieve professional-looking results.

Beam Styles

Faux beams, or box beams, are for decoration and do not actually support a ceiling. The most common design is a simple box made of an inner support plus three pieces that show. The beam may have a rough and rustic look, a sleek appearance, or a cottage style. You can also buy and install ready-made faux beams that may look like an axe-hewn structural support from frontier days.

A Box beams made of rough-sawn cedar add a nice casual touch to this eclectic kitchen. B A coffered ceiling is made of faux beams in a crisscross pattern. To build one, start with straight box beams that span the room. Cut and install shorter beams to span between the long beams, in a repeating sequence. Then add trim pieces to the ceiling and perhaps to the bottoms of the beams as well. C These beams look somewhat like I-beams: the bottom and top plates are wider than the horizontally oriented sides for a modern, casual look. D These are simple box beams with crown molding attached at their tops. A board the width of the beams runs around the room to complete the look.

C

D

Building Box Beams

In the design shown on the following pages, the bottom beam piece is a 1×4; the sides are 1×6s. That makes for a squarish beam that is 5 in. wide and 5½ in. tall. You may prefer to vary the dimensions.

1 **LAY OUT THE BEAMS. Plan for a series of evenly spaced beams. For each beam, measure and mark for both sides of the inside nailing block on each side of the room (top). Working with a helper, snap chalklines between the marks (above).**

2 **MAKE A NUMBER OF PLYWOOD NAILING BLOCKS.** The blocks are as wide as the bottom beam piece, and at the right height so they will hold the bottom piece at the desired height. In this case, ¾-in. plywood is cut to 2 in. wide for the top, and the sides are 4½ in. wide, so that the bottom beam piece will be raised ¼ in. above the bottom of the sides.

3 **ATTACH THE NAILING BLOCKS.** If the beam will run across ceiling joists, use a stud finder to locate the joists and drive 3-in. screws to attach the nailer to the joists. Install a nailing block every 2 ft. or so.

Attaching a Beam with Toggle Bolts

If the beam runs parallel to the ceiling joists there will be no framing to attach it to. Plastic drywall anchors will not be strong enough, so use 3-in. toggle bolts instead.

1. Hold the nailing block in place and drill a hole of the recommended size through it and through the drywall.

2. Thread the toggle nut onto the end of the bolt and tap it through the hole until you can feel the wings of the nut open.

3. Screw the bolt tight; you may need to pull down on it as you drive the bolt.

TOGGLE BOLTS.

1. DRILL THE HOLE.

2. INSERT THE TOGGLE.

3. TIGHTEN THE TOGGLE.

4 **ATTACH THE SIDES.** Cut the side pieces, hold them up against the ceiling, and attach them to the nailing blocks with nails or finish screws.

5 **ATTACH THE BOTTOM PIECE.** Shoehorn the bottom piece into place and tap it up until it hits the nailing blocks. Drive screws or nails up into the box and through the sides.

6 **TRIM OUT THE BEAM.** If desired, add trim where the beam meets the ceiling. Trim options include cove, quarter round, bead molding, simple square strips, or small crown.

Mantel and Fireplace Surround Styles

The styles shown here can be built using the basic instructions on pp. 208–215, then layering on various types of moldings.

A This mantle is somewhat unusual, with its horizontally oriented bands rather than outward-angled crown-molding support. Fancy corbels like these are available from a number of sources B Quality workmanship shines through in the fine lines and detail of this mantel. C Simple geometric lines in broad strokes give this surround Craftsman appeal. When a surround gets this close to the fire, it should be built of nonflammable materials like PVC sheets.

D **Though the total effect is neat and pristine, three unusual trim pieces make this fireplace surround rich in detail. Fluted boards for the pillars and a dentil strip for the mantel can be purchased online. Curly molding pieces for the frieze are also available; you may have to size the frieze board to match available molding.** **E** **Here, trim made with strips of 1×s or rip-cut ¾-in. MDF add depth to the pillars. A curved frieze board, which can be cut out of MDF layers, adds a graceful touch.**

Building a Fireplace Surround

As you design a fireplace, follow local fire codes to ensure that it will be safe to use. In most locales, no wood surface may be within 6 in. of a wood-burning fireplace opening. If the wood protrudes more than a few inches outward, that setback dimension will probably increase. For gas fireplaces or fireplaces with glass inserts, the requirements may be less stringent.

Some terminology: The word *mantel* usually refers to the horizontal shelf above a fireplace, although it sometimes refers to the entire fireplace surround. At the top is the *shelf,* which is what it sounds like. A *frieze* board is horizontally oriented under the shelf; it may or may not contain decorative elements. Supporting the mantel are the *legs,* or *pillars,* on each side of the fireplace.

MANTEL CUTAWAY

This mantel is made of 3/4-in. plywood and three types of molding: plycap, crown, and 3-in. lattice.

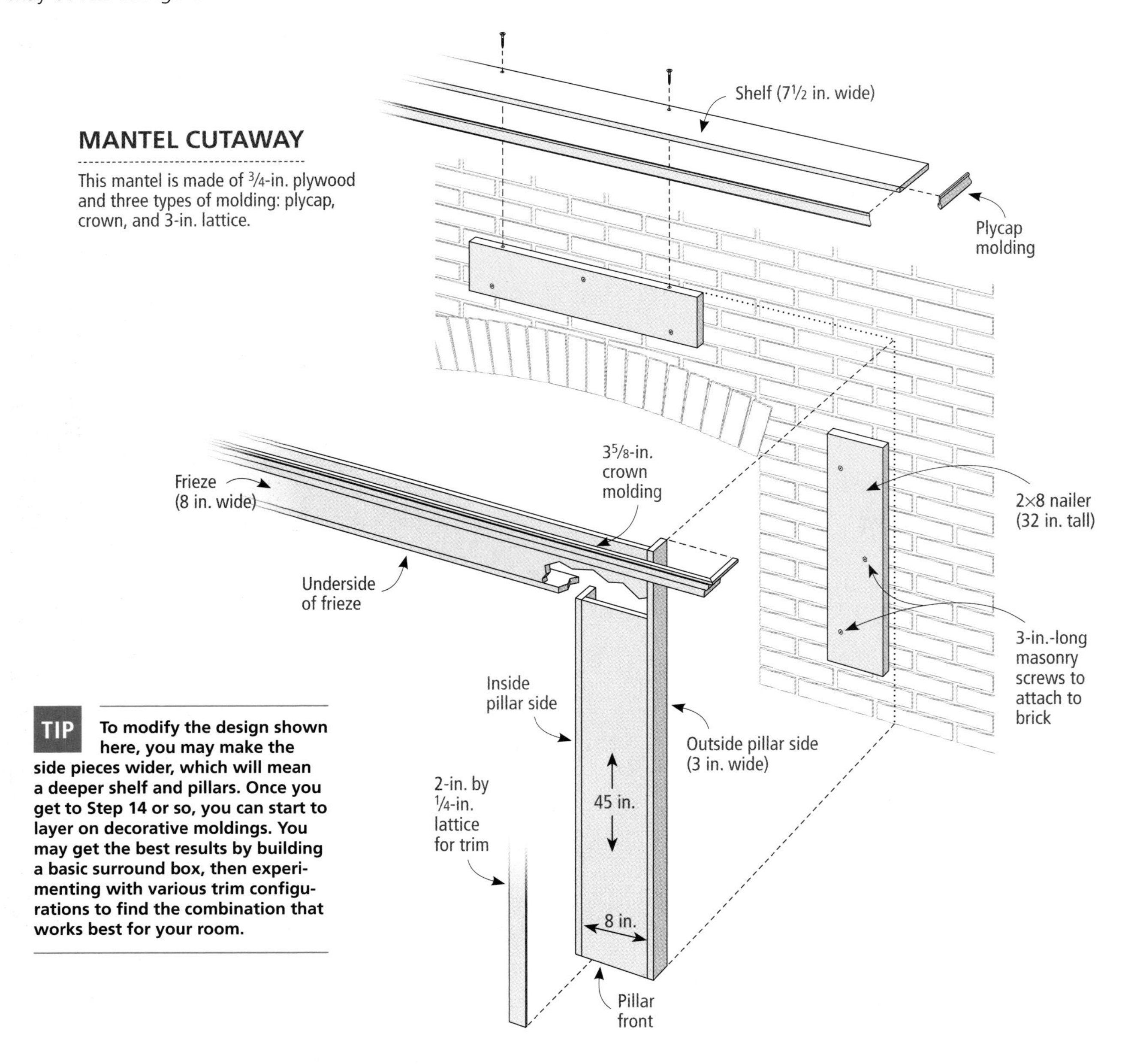

TIP **To modify the design shown here, you may make the side pieces wider, which will mean a deeper shelf and pillars. Once you get to Step 14 or so, you can start to layer on decorative moldings. You may get the best results by building a basic surround box, then experimenting with various trim configurations to find the combination that works best for your room.**

1 **PREPARE THE STOCK.** Rip-cut strips of ¾-in. birch or other hardwood plywood for the shelf, two pillar fronts, four pillar sides, and the underside of the frieze. Do not cut them to length yet, and do not rip-cut the frieze yet. If the wall you are installing onto is not flat, you may have to cut the pillar sides at different widths to compensate for the difference, or you may need to cut notches in them. Cut the pillar front and the two inside pillar side pieces to the desired inside height of the surround.

2 **CROSSCUTTING A WIDE BOARD.** To crosscut a wide board with a miter saw, make a first cut as wide as possible, then flip the board over, lower the blade, and visually align the blade with the cut.

3 **ASSEMBLE THE PILLARS.** Apply glue and drive nails to attach the inside sides to the pillar fronts, keeping their edges flush.

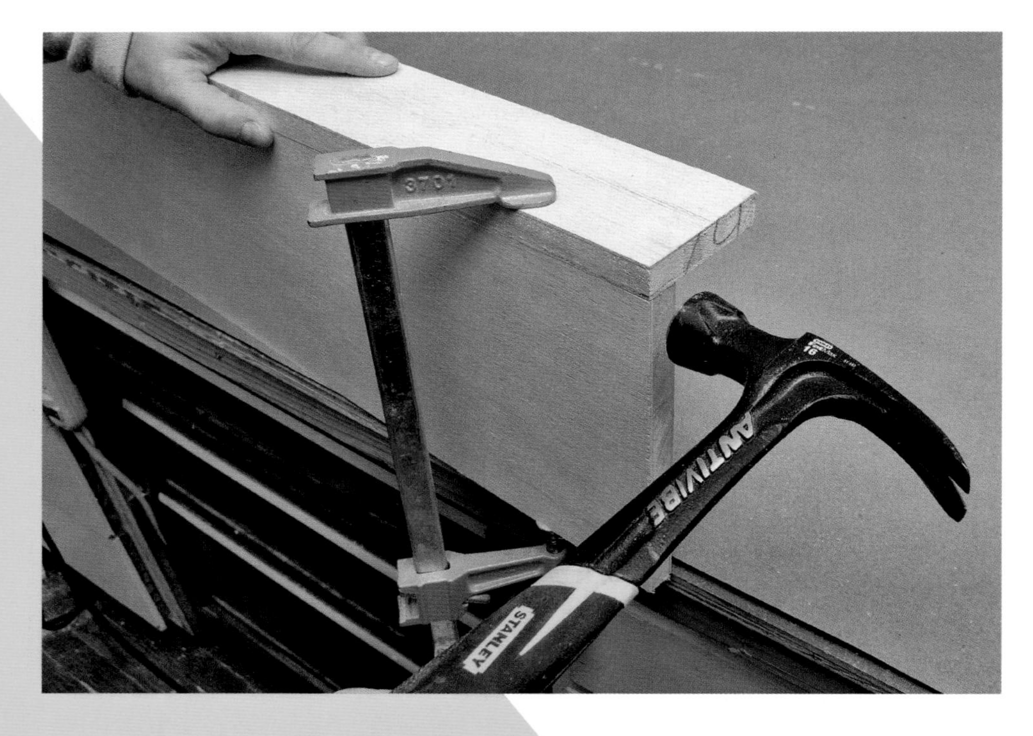

Clamp and Tap to Keep It Flush

When joining the pillar and frieze assemblies, keep the ends, sides, and faces all precisely flush for a neat finished appearance. If it is difficult to hold the pieces perfectly flush and still, clamp them together with medium pressure and tap with a hammer to get the position just right. Then drive nails.

4 **ATTACH THE OUTSIDE PILLAR SIDES.** Cut the outside pillar side pieces so they are as long as the pillar fronts plus the desired height of the frieze (including the ¾-in. thickness of the frieze's bottom piece). Attach one of them with glue and nails so the bottom is flush with the bottom of the pillar and the top extends upward (left). Place the two pillars side by side to be sure you position the other outside piece on the correct side, and attach it as well (right).

TIP For a symmetrical look, the shelf should overhang the frieze by the same amount on the sides as in front. As you measure, don't forget to factor in the 3⁄4-in.-thick side pieces that will be on each side of the frieze.

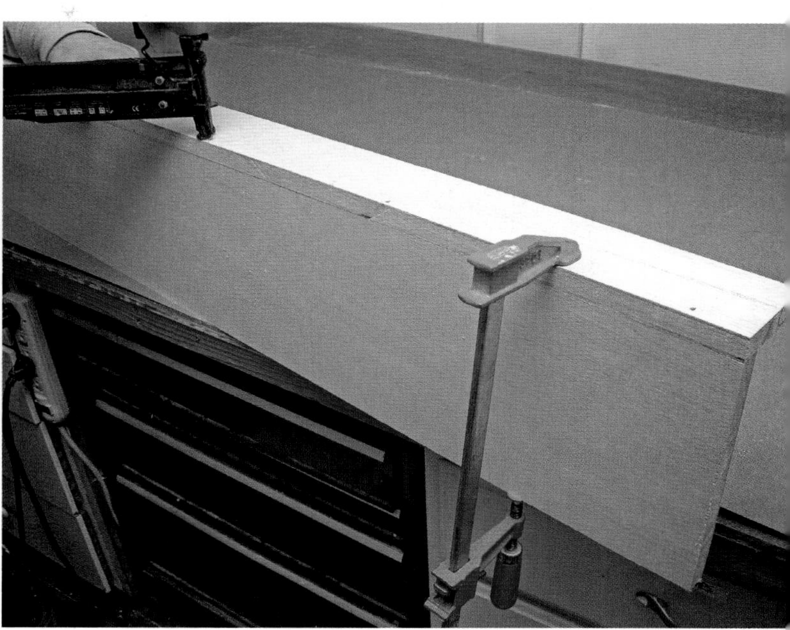

5 **MEASURE FOR THE FRIEZE.** To find the frieze's width, measure the overhang of the pillar sides (top) and subtract 3⁄4 in. To find the frieze's length, put the two pillars together, measure between the outside pieces (above), and add the desired inside width of the fireplace surround. Cut the frieze and the frieze's bottom piece to the same dimension.

6 **ASSEMBLE THE FRIEZE.** Attach the frieze's bottom piece with glue and nails.

7 **ATTACH THE SHELF.** Cut the shelf to the desired length. Place the frieze assembly on a block that is 3⁄4 in. narrower than the side piece so it is level on the table. Measure and draw a line for where you will drive nails through the top of the shelf and into the center of the frieze board (left). Place the shelf so it overhangs the same distance on each side and drive finish nails to attach it to the frieze (right).

8 **POSITION THE SURROUND.** Working with a helper, set the pillars and the mantel in place. Check for a level shelf and pillars that are square to the mantel, and measure to see that the pillars are the same distance from the firebox on each side. Mark reference lines for the pillars, or just hold them tightly in place as you remove the mantel for the next step.

9 **MARK FOR THE NAILERS.** Mark the wall for the insides of the pillars, where you will put the nailers. (The 2×8 nailers we use are about 1 in. narrower than the pillars' inside dimension, which allows some leeway.)

Notching around Base Molding

If the pillars bump into the wall's base molding at the bottom, you could cut back the base molding. Or notch the pillar, as shown here. (Here we show a simple base shoe, but the same method works for baseboard.)

1. Use scissors or a knife to cut a cardboard template, and test that it fits against the molding.
2. Use the template to scribe a line on the pillar (**2A**), and cut with a jigsaw (**2B**).

1. MAKE A TEMPLATE.

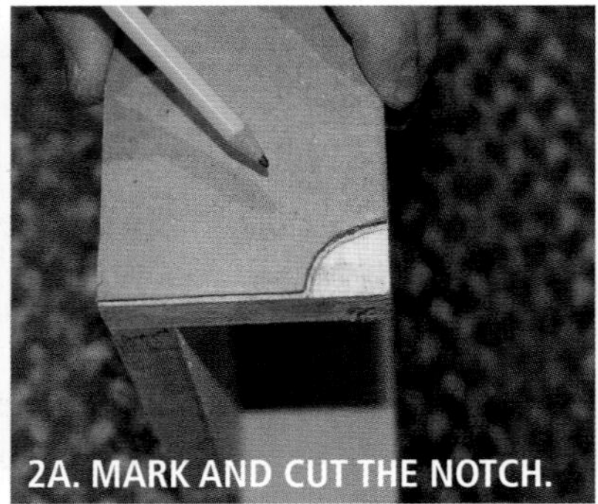

2A. MARK AND CUT THE NOTCH.

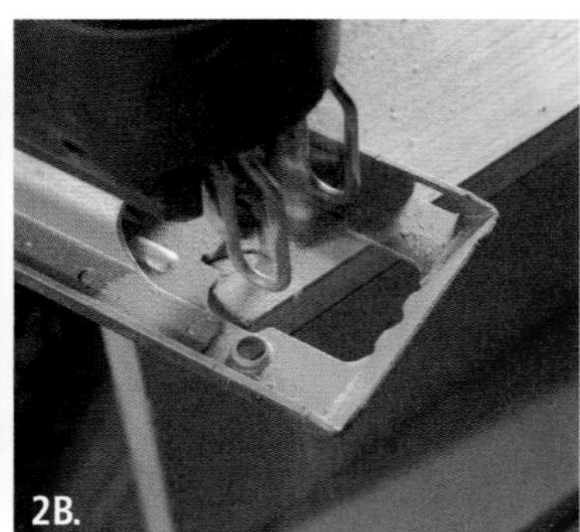

2B.

TIP If you have a regular stud wall around the fireplace, you can simply drive wood screws to attach the nailers.

10 **PREPARE TO ATTACH THE SIDE NAILER.** If you have a brick wall, use masonry screws, which attach firmly when driven into a hole that is drilled using the correct matching masonry bit (above). Cut a piece of 2× lumber to fill most of the area behind the pillar. Drill three or more holes for attaching a nailer. Hold the nailer plumb and against the marks from Step 9, and start to drive a hole into the brick (right).

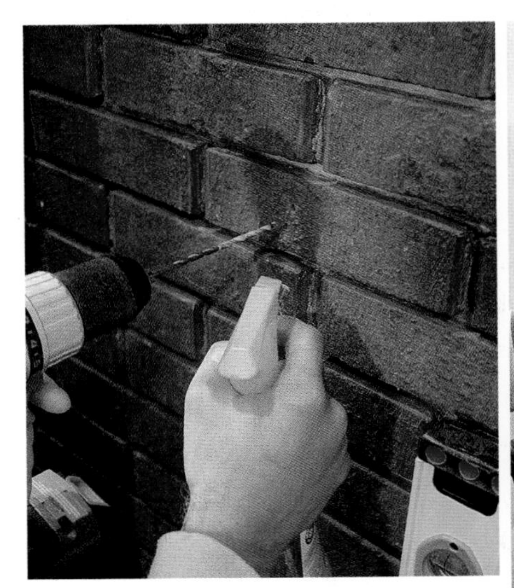

11 **KEEP DRILLING.** Remove the nailer and continue drilling the hole. If the bit heats up, it can quickly become dull, so every so often pull the bit out and squirt with a soapy solution or water to keep it cool.

12 **ATTACH THE SIDE NAILER.** Drive the screw to hold the board in place (left), then check for plumb (right) and partially drive the other holes to mark their locations. Remove the screw and the nailer, finish drilling the other two holes, and reattach the nailer with all three screws.

13 **ATTACH THE TOP NAILER.** Mark for the top nailer behind the frieze board, and attach it as well.

TIP For this project, we installed the crown molding upside down—with the detail at the top rather than the bottom, because that makes it extend farther outward to better support the shelf.

14 **ATTACH THE SURROUND.** Assemble the pillars and mantel over the nailers. Drive finish-head screws to attach them to the nailers. Use a scrap of the lattice molding you plan to attach as a guide to be sure all the screw holes will be covered with trim. Fasten to all the nailers with screws, and also fasten the outside side pieces to the mantel using finish nails.

15 **ATTACH THE SHELF MOLDING.** Cut plycap molding to wrap around the front edge of the shelf, with 45-degree bevel cuts at the corners. Fasten with glue and pin nails or small finish nails.

16 **INSTALL THE CROWN.** Cut crown molding, as shown on p. 158, and attach with nails.

17 **DRESS UP WITH LATTICE TRIM.** Cut 2-in. lattice pieces and attach with glue and pin nails, first for the vertical outside pillar edges, then for the horizontal piece spanning between them, and then for the vertical inside pillar pieces. Cut short pieces to fit between the long pieces, and experiment with designs until you come up with one you like.

18 **ADD A KEYSTONE.** For this mantel a "keystone" piece cut from ¼-in. plywood adds some style at the center of the frieze.

19 **FILL AND SAND.** Fill the nail holes with wood filler or spackle and sand smooth when dry. You may choose to soften the lattice edges by sanding them a bit.

20 **THE FINISHED PROJECT.** Prime and paint. If you like, add a decorative feature in the center of the frieze.

INDEX

CREDITS

All photos by Steve Cory and Diane Slavik, except as noted below:

p. 2: IP Moulding

p. 3: Deep River Partners

pp. 4–5: (A, C, E) Deborah Leamann Interior Design, (B) Foran Interior Design, (D) Focal Point Products, (F) Bill Freeman, Celebration Development Group, LLC

pp. 6–7: (A) Glidden Company, a division of PPG Industries, (B) Amy Etra, photographer; design by Bill Freeman, Celebration Development Group, LLC, (C) Foran Interior Design, (D) Acanthus Architecture & Design, (E) Tech Lighting

pp. 8–9: (A) Acanthus Architecture & Design, (B) IP Moulding, (C) John Wilbanks, photographer; design by Kathryn Tegreene Interior Design, (D) Matthew Millman, photographer; design by Schwartz and Architecture, (E) Bill Freeman, Celebration Development Group, LLC, (F) David Duncan Livingston, photographer; interior design by Adeeni Design Group

pp. 10–11: (A) Lisa Wolfe Design, Ltd., (B, E) IP Moulding, (C) Deborah Leamann Interior Design, (D) Ambient Ideas Photography, Dreamstime

pp. 12–13: (A, B) IP Moulding, (C) Matthew Millman, photographer; design by Schwartz and Architecture, (D) John Wollwerth, Dreamstime, (E) Anthony Berenyi, Dreamstime

pp. 14–15: (A) Larry Malvin Photography, Dreamstime, (B) Deborah Leamann Interior Design, (C) Focal Point Products, (D) The Kohler Company

pp. 16–17: (A) Acanthus Architecture & Design, (B) Focal Point Products, (C, E) Tech Lighting, (D) YinYang, iStock

p. 18: Metrie

p. 81: Glidden Company, a division of PPG Industries

p. 97: Armstrong World Industries, Inc.

p. 120: Metrie

p. 121: The Sherwin-Williams Company

p. 130: Metrie

p. 147: The Sherwin-Williams Company

pp. 148–149: (A) The Dutch Boy Group, (B) Glidden Company, a division of PPG Industries, (C, F) Tech Lighting, (D) Focal Point Products, (E) John Wollwerth, iStock, (G) photodiva/travelfotostock, iStock

p. 170: Amy Etra, photographer; design by Bill Freeman, Celebration Development Group, LLC

p. 171: Metrie

pp. 172–173: (A) Glidden Company, a division of PPG Industries, (B, C, D) Armstrong World Industries, Inc., (E) The Dutch Boy Group, (F) Winder Gibson Architects

p. 198: Michael Del Piero Good Design

p. 199: Irina88w, Dreamstime

pp. 200–201: (A) Tech Lighting, (B, C) Armstrong World Industries, Inc., (D) Irina88w, Dreamstime

pp. 206–207: (A) Focal Point Products, (B) Winder Gibson Architects, (C) Tech Lighting, (D, E) Glidden Company, a division of PPG Industries

STANLEY

Publications

NEW
Expert advice for DIYers

MASTER EVERY PROJECT

OPEN A TOOLBOX OF PROFESSIONAL ADVICE with the ***STANLEY Quick Guide*** series. Highly visual and easy to use, these laminated, spiral-bound guides are packed with how-to photographs and no-nonsense instructions. Now you can tackle home projects with confidence.

EASY HOME REPAIRS

Spiral
EAN: 9781631861642
5 x 8, 32 Pages
Product #083041
$7.95

GARAGE STORAGE SOLUTIONS

Spiral
EAN: 9781631861635
5 x 8, 32 Pages
Product #083040
$7.95

HOME ENERGY SAVINGS

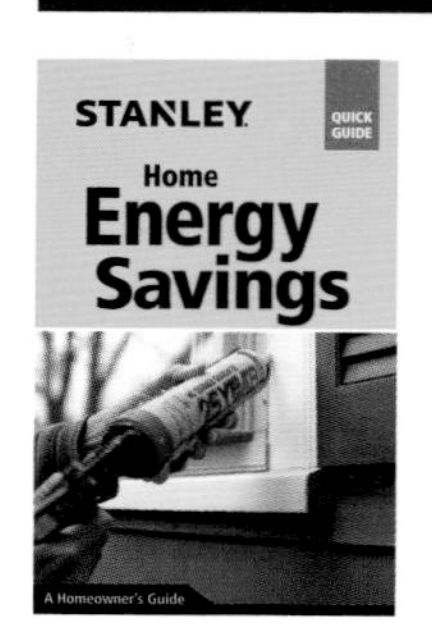

Spiral
EAN: 9781631860034
5 X 8, 32 Pages
Product #083034
$7.95

EASY HOME PLUMBING REPAIRS

Spiral
EAN: 9781627109857
5 X 8, 32 Pages
Product #083031
$7.95

EASY HOME WIRING REPAIRS

Spiral
EAN: 9781631860027
5 X 8, 32 Pages
Product #083033
$7.95

EASY HOME DRYWALL REPAIRS

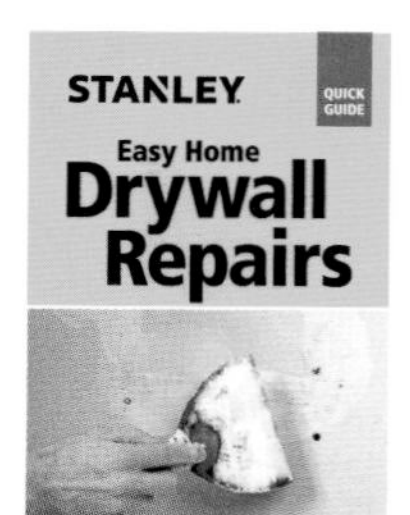

Spiral
EAN: 9781627109840
5 X 8, 32 Pages
Product #083030
$7.95

© 2015 The Taunton Press

Available at TauntonStore.com or wherever books are sold.

STANLEY and the STANLEY Logo are trademarks of Stanley Black & Decker, Inc. or one of its affiliates, and are used under license.